HOW FAR IS UP?

HOW FAR IS UP?

MEASURING THE SIZE OF THE UNIVERSE

JOHN AND MARY GRIBBIN

Icon Books

Published in the UK in 2003
by Icon Books Ltd., Grange Road,
Duxford, Cambridge CB2 4QF
E-mail: info@iconbooks.co.uk
www.iconbooks.co.uk

Published in Australia in 2003
by Allen & Unwin Pty. Ltd.,
PO Box 8500,
83 Alexander Street,
Crows Nest, NSW 2065

Sold in the UK, Europe, South Africa
and Asia by Faber and Faber Ltd.,
3 Queen Square, London WC1N 3AU
or their agents

Distributed in Canada by
Penguin Books Canada,
10 Alcorn Avenue, Suite 300,
Toronto, Ontario M4V 3B2

Distributed in the UK, Europe,
South Africa and Asia by
Macmillan Distribution Ltd.,
Houndmills, Basingstoke RG21 6XS

ISBN 1 84046 439 9

Typeset by Cambridge Photosetting Services

Printed and bound in the UK by
Mackays of Chatham plc

For Jon and Ben

CONTENTS

CONTENTS

ILLUSTRATIONS

INTRODUCTION

THERE'S a trick question that goes 'How far can you see on a clear day?' The trick is that on a clear day you can always see the Sun, which is 150 million km away. We are used to thinking about distances we can travel on Earth, where even if we flew right around the globe we would cover a distance of only about 40,000 km. But we are not used to thinking of distance upwards. Few people have any idea how far 'up' goes, and it wasn't until well into the twentieth century that astronomers began to grasp the enormous scale of the Universe. The Ancients believed that the stars and planets were tiny lights attached to crystal spheres nested inside one another and rotating around the Earth. The scale of these spheres would have been comparable to the terrestrial scale of things, with diameters of perhaps a few thousand kilometres. Even after Nicolaus Copernicus published his sensational suggestion that the Earth moves around the Sun in 1543, many people tried to cling to the old ideas, with the stars imagined fixed to a single crystal sphere just outside the orbit of the most distant planet from the Sun, and each planet attached to its own crystal sphere. When Galileo first turned a telescope skyward early in the seventeenth century, and saw that the band of light across the sky known as the Milky Way is made up of a myriad of individual stars, too faint to be picked out by the naked eye, for those with eyes to see it became clear that the Universe must be enormously bigger than the Solar System, and that the stars must be other suns, millions of times further away from us than the Sun itself.

But these ideas were taken on board only by the cognoscenti. The image of the crystal spheres was only truly shattered in the seventeenth and eighteenth centuries, when the work of astronomers such as Edmund Halley first showed that the orbits of comets pass right through the region where the planetary crystal spheres were supposed to be, and then that the stars themselves move independently, and are not fixed to anything. By that time, though, astronomers already knew that they were dealing with distances vastly greater than the circumference of the Earth. As long ago as 1671, using the standard surveying technique of triangulation, two teams of French observers, one making measurements from Paris and the other from Cayenne in French Guiana, determined the distance to the Sun as 140 million km – only about 10 per cent less than the best modern estimate.

So, how far away is the nearest star that you can see in the sky? It wasn't until the nineteenth century that

astronomical techniques became good enough to measure accurately the distances to even the nearest stars, but
again they did so by triangulation. Measuring the
distance to the Sun by triangulation involved a baseline
stretching from Paris to Cayenne; measuring the distances to the nearest stars involves a baseline stretching
over the width of the Earth's orbit around the Sun, 300
million km, making use of observations carried out six
months apart when the Earth is on opposite sides of the
Sun. The results were stunning. People have travelled
384,400 km to walk on the Moon, but the furthest stars
you can see could never be reached in anyone's lifetime.

In order to understand such distances at all, we need a
new way to measure them; writing out the distances in
kilometres is just ridiculous. The new way of measuring
the Universe involves light, which always travels at a constant speed of 300,000 km per second. So, it takes light a
little over one and a quarter seconds (almost exactly 1.28
seconds) to travel to the Moon. Because radio waves
travel at the speed of light, when ground control on Earth
speaks to astronauts on the Moon, there will always be a
delay of at least 2.5 seconds before the reply reaches the
Earth, for the round-trip journey of the radio waves. So
the Moon can be said to be 1.28 'light seconds' away from
Earth. Similarly, a light year is the distance light can move
in one single year. Light moving at 300,000 km per second can travel 9.5 thousand billion kilometres in a year.
This is a mind-boggling distance, but even the nearest
star to the Sun is 4.3 light years (that is, just under 41
thousand billion km) away. The far-off stars, the other
suns that you see as distant specks of light high up in the

dark sky, are tens of light years – some of them hundreds of light years – away from us. But this is only the beginning of the story.

In the twentieth century, as bigger telescopes were built and more sensitive instruments were developed to hang on the ends of those telescopes, it became clear that everything we can see with the naked eye is just a tiny local region of the Universe at large. The first crucial steps out into this vast Universe were taken by an arrogant, opinionated American astronomer, aided by a man with no formal education to speak of who had once made his living driving mule trains. The story was completed, at the end of the twentieth century, by large teams of astronomers using the most sophisticated telescope built up to that time, the orbiting space telescope named after the pioneer Edwin Hubble.

On a clear night you can see, with your own eyes, amazingly much further than you can on a clear day; but even that is near compared to the incredible distances away across the Universe that can be seen by the Hubble Space Telescope. This book tells of the twentieth-century quest to look as far out into the Universe as possible, and to measure accurately the distance between Earth and the furthest objects that can be detected by telescopes. People have always looked up into the sky on starry nights and longed to know more about the blanket of stars surrounding them. This longing drove people to develop ways and means of finding out more, and this story tells how people worked towards really understanding how far is 'up'.

1

THE BOY ON THE MOUNTAIN

NINETEEN-hundred-and-five was a big year in science. It was the year in which Albert Einstein published his Special Theory of Relativity, which described the behaviour of objects moving at very high speeds. But that didn't have much significance for a fourteen-year-old boy called Milton Humason, who was heading for a summer camp high up on Mount Wilson near Pasadena, just north of Los Angeles, in California. He loved to look up at the stars at night but wasn't remotely academic and never thought about studying them at all. That summer on Mount Wilson was the best time that Milton could remember, and he dreaded going back home to school. But a remarkable thing happened. Back in 1905, it was legal for young people to be working full-time when they were fourteen, and Milton was so obviously unhappy when he was back home studying that his parents agreed to let him take a year out of school to take up a job that he had been offered in the Mount

Wilson Hotel as a bellhop, running errands for hotel guests, and part-time handyman. Milton Humason's parents hoped that a year of constant hard work up a mountain would send Milton gratefully back to college.

But their plan didn't work. Milton Humason loved Mount Wilson so much that he never went back to school or had any more formal education. Discovery is often down to being in the right place at the right time, and Milton's decision to stay on Mount Wilson led him, within 25 years, to play a major part in showing that Albert Einstein's General Theory of Relativity (which goes far beyond the Special Theory, and explains how gravity works) is a good description of the Universe we live in. During his lifetime, Milton worked with Edwin Hubble, the man who discovered that the Universe is expanding and after whom the famous Hubble Space Telescope is named. He learned so much from his own observations and those of the people he worked with that he went on to teach Allan Sandage – the man who went on to measure how fast the Universe is expanding – how to use large astronomical telescopes.

The boy who had chosen to work on Mount Wilson simply because he loved the outdoor mountain life arrived there at a time when Mount Wilson was about to become one of the most important centres of astronomical research in the world. At the start of the twentieth century, astronomy was beginning to become big science because the technology had been developed to make huge, accurate telescopes that could look far up into the Universe. They were the equivalent in their time of the giant 'atom smasher' machines at laboratories like

6

CERN, in Geneva, today. But this kind of observational astronomy was expensive – not just because the new telescopes were vast and used the latest technological developments. Observatories cost an enormous amount of money to construct because they had to be built on mountain-tops high above the pollution of city life and as far above the obscuring effects of the Earth's atmosphere as possible. And it was money that helped George Ellery Hale to bring astronomy into the twentieth century. He had become Professor of Astronomy at the University of Chicago in 1892 at the age of 24. He had an excellent university education at MIT (Massachusetts Institute of Technology), a brilliant brain and an extremely rich father.

Hale was undoubtedly a good astronomer, but not all astronomers can raise the money to build a telescope. Because he had such a wealthy family, George Ellery Hale had met a lot of rich people in his lifetime and he knew how to make the most of his contacts. Mount Wilson Observatory was built because Hale knew who to ask for the cash. He started in Chicago, raising funds for a new observatory centred around a refracting telescope (the kind that uses lenses) with a main lens 40 inches, or just over a metre, across, which is still the biggest telescope of its kind in use today. Hale knew that in order to study fainter objects in the sky he would need even larger telescopes on high mountains. The 40-inch was just about as big as a refractor can be, because above this diameter the weight of the lens makes it bend and distorts the view that can be seen through it. Bigger telescopes would have to be reflectors, using mirrors that could be supported from behind without interfering with the incoming light from distant stars which reflects off the front of the mirror.

In 1903, George Ellery Hale went on a reconnaissance trip up into the mountains in California looking for a site on which to build a new observatory. He spent that summer in an abandoned log cabin on Mount Wilson, making observations of the night skies with a small portable telescope to check just how clear the stars above the mountain were. That summer of clear skies convinced him that Mount Wilson was the place on which to build the biggest telescope in the world. There is an old saying that money makes money. Hale's rich father had already bought a 60-inch-diameter (about 1.5-metre) mirror back in 1896 for his son to use to help in the construction of what he hoped would one day be the biggest telescope the world had ever known. It was probably the fact that he already owned this mirror that enabled George Ellery Hale to persuade the Carnegie Institution in Washington to put up $150,000 (a huge amount of money in those days) to build an observatory on the mountain.

Work had begun on the construction of the Mount Wilson Observatory in 1904, a year before Milton Humason first visited the mountain. But building on top of a mountain is never easy. When Hale first went up Mount Wilson back in 1903 he had used an old Indian trail, winding 8 miles round the mountain, to get to the

Illustration 1. Comparison of refracting and reflecting telescopes. (Illustration copyright © 2003 Nicholas Halliday)

A telescope must gather light from a faint object and then concentrate and magnify it, making it bright enough for the observer to see or for instruments to record. In a 'refracting' telescope, a convex lens is used to gather and focus the light, whereas in a 'reflecting' telescope, the light is gathered by a curved mirror. An eyepiece lens then spreads the light out over a greater portion of the observer's retina, magnifying the image.

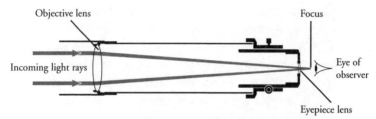

Refracting telescope

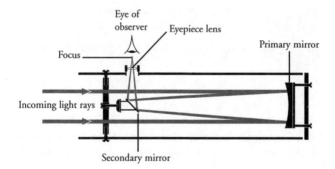

Reflecting telescope

Illustration 2. Andrew Carnegie, the founder of the Carnegie Institution of Washington, with George Ellery Hale in March 1910. As an advisor to the Institution, Hale was instrumental in convincing it to agree, in 1904, to support construction of the 60-inch telescope on Mount Wilson, seen here in the background. (Henry Huntington Library, San Marino, California)

summit. This trail was far too narrow to use to transport building materials; so, before the observatory could be built, a road would have to be constructed to reach the site of the proposed observatory at the top of Mount Wilson, 2,000 metres high. This road, along which all the materials to build the telescopes and living accommodation for astronomers as well as the tents, food and fuel for the construction teams had to be hauled by pack mule, was less than a metre wide. The road that led to the construction of the biggest observatory in the world was just a narrow dirt trail zigzagging to the top of the mountain, along which pack mules struggled and people trudged on foot or on horseback. But this relentless sweat and strain paid off. By hauling the greatest technological developments in astronomy then known to man up a dirt track to the clear air at the top of Mount Wilson, the new reflector with the 60-inch mirror became operational in 1908.

Working away on the lower slopes at the Mount Wilson Hotel, Milton Humason watched all this activity with fascination. At about the time that the 60-inch telescope started to gaze up into the night sky, he left his job at the hotel to become a mule driver running pack trains up Mount Wilson to the construction site at the top. By this time, the trail had been widened enough for a horse-drawn carriage to drive up – but this was only for the very brave! Although the main telescope was in place, there was still plenty of construction going on after 1908. In 1906, George Ellery Hale, who never seemed to be satisfied with what had been accomplished, had persuaded John D. Hooker, a Los Angeles businessman, to put up the money for an even bigger telescope with a main mirror 100 inches (about 2.5 metres) in diameter. In return for his financial help, the instrument would be called (and still is to this day) the Hooker Telescope.

Illustration 3. The 60-inch reflector at Mount Wilson, first used in 1908. According to Allan Sandage, 'The Mount Wilson 60-inch telescope was the granddaddy of them all, where many of the problems of telescope design and solutions were first understood.' (Courtesy of Carnegie Observatories, Carnegie Institution of Washington)

Working high up on a mountain-top as a mule-train driver to build the greatest observatory ever known sounds like a romantic dream to most of us; but to crown

Illustration 4. A horse-drawn 'truck' proceeds up Mount Wilson's thin trail c. 1905. Lighter building materials and supplies for the Observatory were brought up by mule train. (Courtesy of Carnegie Observatories, Carnegie Institution of Washington)

this fantastic adventurous existence, Milton Humason found love. In 1911, when they were both just twenty years old, he married Helen Dowd, the daughter of one of the key workers on the project, Merrit Dowd, who later became the chief electrical engineer on the mountain. For the next two years, Milton carried on working on the mountain he loved so much. But when Helen gave birth to their first child in October 1913, they realised that Milton had to get a better job to support the family and that the time had come to settle down. Love for his family would have to replace love of his mountain, so Milton, Helen and their baby left Mount Wilson and moved down to the town of Pasadena, where Milton found a job as head gardener on an estate.

Illustration 5. Though he started work at Mount Wilson Observatory as only a janitor, Milton Humason's aptitude for observing soon led him to become a first-rate astronomer, whose efforts were essential in the discovery of the expanding universe. (Courtesy of Carnegie Observatories, Carnegie Institution of Washington)

After three years, Milton and Helen Humason had saved enough to buy their own fruit farm – what the Californians liked to call a 'citrus ranch' – outside Pasadena. Life to any outside observer must have looked perfect, but Milton Humason had never really settled to life away from the mountain. When he heard from his father-in-law Merrit Dowd that there was a job going for

14

him at the Mount Wilson Observatory if he wanted it, he couldn't resist. So Milton threw up the prospect of working on his own fruit farm in one of the most lovely areas of Californian farmland to work as one of three janitors looking after the buildings and clearing up around the Observatory.

Being a replacement janitor didn't look like much of a job, but the 100-inch telescope was about to start working and therefore, with more astronomers visiting the mountain and using this and the 60-inch telescope, as well as being a janitor Milton was to be expected to work part-time as a 'night assistant'. It was November 1917 – and the Great War in Europe was still raging – when Milton Humason started work helping out with whatever the astronomers at Mount Wilson Observatory needed help with. He made sure that the telescopes were pointing in the right direction, helped to develop the photographic plates and made the coffee, all for around $80 a month. Even back in 1917, $2.50 a day wasn't enough to raise a family without a struggle; but, as the saying goes, he would willingly have paid them. Milton loved the mountain, he loved the work, and a rent-free cabin went with the job; in addition, all his meals were free while he was working. There is no record of how Helen Humason viewed her husband's change of career.

At the time Milton started work at the Mount Wilson Observatory, the only way in which astronomers could record images of stars and other objects in the sky was by using glass photographic plates – pieces of glass covered in chemicals sensitive to light. Even with the large telescopes then in use on Mount Wilson, most of the objects the astronomers were interested in studying were so faint that the plates had to be exposed to the light being focused by the telescope for hours, while the observer had to keep the telescope steadily pointing at the object in the sky that they were concentrating on. The Hooker telescope did have an automatic clockwork system for guiding it to track a particular star as the night progressed, but this was very unreliable and needed constant re-adjusting throughout the night. In some cases, not even a whole night's observing would be long enough to take a single photograph. When this happened, the observer had to pack the glass photographic plate away in a dark box, then take it out the next night, remount it carefully in the telescope in exactly the same position as before and allow it to absorb more light from the faint object up in the sky which was being recorded. This further night of observing would bring out the photographic image more clearly.

Only after all that work – sometimes several consecutive nights of trying – could the plate be taken down to the photographic dark room and developed by treatment with more chemicals to fix the image on the plate, as a photographic negative on which bright stars showed up black. Milton Humason was bewitched by the whole process and longed to know more. His wish was granted when Hugo Benioff, a university student from Pasadena, came up to the Mount Wilson Observatory to work as a volunteer in his vacation. Hugo liked Milton and took the time to teach him how to take astronomical photographs, using a little 10-inch telescope. The patience that Milton must have developed in his years as a mule-train driver,

Illustration 6. Funded by the businessman John D. Hooker, the 100-inch telescope became operational in November 1917. Its mirror remains the largest solid plate glass mirror ever made. (Royal Astronomical Society)

coupled with a simple knack for the job and interest in the subject, encouraged him to carry on taking astronomical photographs even after Hugo Benioff had returned to college.

Soon Harlow Shapley, one of the senior astronomers on Mount Wilson, noticed Milton's talent and took him on as his unofficial assistant. Milton Humason, ex-bellhop, handyman, gardener, citrus farmer and janitor with no college education, soon became in Harlow Shapley's words 'one of the best observers we have ever had'. Shapley had so much confidence in Milton's ability that he went to see George Ellery Hale, who was still the Director of the Mount Wilson Observatory, to urge him to officially appoint Milton Humason onto the scientific staff at the Observatory. Hale, although scepti-cal that Milton Humason, a man with no academic attainments, was the man for the job, trusted Harlow Shapley's judgement enough to appoint Milton officially as a junior member of the scientific staff of the Observatory in 1920, promoting him to Assistant Astronomer in 1922.

By the time that Milton Humason received that key promotion, Harlow Shapley had left Mount Wilson to work as Director of the Harvard Laboratory. But just before he left Mount Wilson, he had been involved in a strange incident in which his actions may well have held back the discovery of the expanding Universe for several years. This happened in the winter of 1920/1 after Harlow Shapley had told Milton to take some photographic plates of an object known as the Andromeda Nebula – a faint patch of light in the sky – using the 100-inch tele-scope, which had been operational since 1918. Shapley asked Milton to take several pictures of the nebula at dif-ferent times and to compare them to see if anything had changed. At the time, it was widely thought that these nebulae might be clouds of gas inside the Milky Way, and so relatively near to us; it was natural to look to see if there were any signs that the 'gas clouds' might be rotating or changing in some other way.

Milton Humason got on with the job as he was asked, but he was surprised to find that in his best images of the Andromeda Nebula, obtained with the best telescope on Earth at that time, there were tiny spots of light (actually black spots, on the negative plates) visible on some plates, but not on others. They looked to him like stars that were varying in brightness, so that sometimes they could just be detected, while at other times they were invisible. Milton carefully marked the positions of these stars with little lines drawn in ink on the back of one of the plates, and took it to Harlow Shapley to show him what he had found. In a staggering lapse of open-mindedness, Shapley calmly explained to Milton that it was well known among astronomers that it was impossible for variable stars to exist inside the Andromeda Nebula. He then took the plates out of Milton Humason's hand, turned them over and wiped the carefully marked ink lines away using a clean handkerchief.

Milton thought much but said nothing. He was well aware that he had only just got his foot on the first rung of the ladder of a serious career in astronomy. He had no academic qualifications, owed his job to Harlow Shapley and was hoping for promotion. Despite his own convictions, he wasn't going to argue with the boss. But Milton's observations had been correct, and within a decade it was established beyond doubt that there are variable stars in the Andromeda Nebula, and in many other similar nebulae that show up as fuzzy blobs on astronomers' photographic plates. These nebulae were not clouds of gas between the stars of the Milky Way at all but whole galaxies like the Milky Way itself – but far, far beyond it. Edwin Hubble, the person who was later to prove this, was already on Mount Wilson when, unknown to him, Harlow Shapley was elsewhere in the same observatory calmly rubbing out the first real evidence that nebulae are galaxies from Milton Humason's photographic plates.

Only after the true nature of galaxies was established did Milton Humason come clean and tell his colleagues just how close Harlow Shapley had been to discovering the truth in 1921.

2

THE MAN WHO LOOKED
BEYOND THE MILKY WAY

THE names of Edwin Hubble and Milton Humason
will always be linked in the history of astronomy for
the work that they did together after 1926 which proved
that the Universe is expanding and implied that it was
born in a 'Big Bang'. They collaborated very successfully
even though they came from very different backgrounds
and possessed dramatically different personalities.

Edwin Hubble was born in 1889, making him two years
older than Humason. Many accounts of his life
suggest that he was something of a superhuman, an all-
round athlete of international class, a boxer who could
have been heavyweight champion of the world, a suc-
cessful lawyer who gave up a brilliant career to be an
astronomer, and a war hero wounded in France in the last
days of World War I. Unfortunately all these stories are a
bit exaggerated, mainly because they are based on Edwin
Hubble's own accounts of his achievements. The truth is
that he went to university in Chicago where he became

a good college athlete (but never world class at anything) and was indeed a first-class student. As well as studying physics and mathematics, he was a student of classics and political economics, winning a coveted and prestigious Rhodes Scholarship which enabled him to study law for two years at the University of Oxford.

It was typical of Edwin Hubble, though, to add a little touch of fantasy to his everyday life. In the two years in which he was studying at Oxford, he turned himself into a kind of counterfeit English gentleman. He took to wearing tweed jackets, smoking a pipe and speaking with a 'British' accent that got right up the nose of many of his American colleagues.

In 1913, Edwin Hubble's father died at the young age of 52, leaving eight children, the youngest of whom were still at school. Edwin couldn't go home until he had finished his studies, but then went back to sort out his family's affairs. Rather than practising law (as he always claimed), he worked for a year as a high-school teacher before going off to Yerkes Observatory near Chicago (the first observatory that George Ellery Hale had founded) to become a research student in astronomy. He completed his PhD in 1917, mainly for his work on a photographic survey of the faint nebulae. He had already been offered a job at Mount Wilson, where Hale was building up his team ready for the start of operations with the 100-inch Hooker telescope. But it was April 1917, the United States had just entered the Great War, and Edwin Hubble volunteered for the infantry, asking Hale to keep the job open for him until he returned from Europe.

In fact, Hubble's infantry division reached France only

in the last weeks of the war. Official records say that he never saw action and give no mention of a wound, although for the rest of his life Edwin Hubble maintained that he couldn't straighten his right elbow properly because he had been hit by shell fragments in battle. At the end of the war, he managed to linger long enough in England to annoy Hale, who wanted him to start work at the 100-inch. He only arrived back in California to take up his new job in September 1919, almost a year after the Great War in Europe had ended. At 29 years old, Hubble continued to use his military title of Major in spite of his very limited army career. But although he annoyed his colleagues with his arrogance and unceasing ability to present himself in the best possible light, Edwin Hubble was undoubtedly one of the best observational astronomers in the world – but not quite as good as Milton Humason.

It is no coincidence that Harlow Shapley and Edwin Hubble were both interested in nebulae. These objects were the big puzzle in astronomy at the time, because finding out just what the nebulae were, and where they were, would have a big impact on our understanding of our own place in the Universe. You have to remember that it was only in the twentieth century that astronomers first began to get a grip on how big the Milky Way itself is. It is very difficult to measure distances to stars, simply because they are so far away. In the nineteenth century, just a handful of star distances had been calculated directly, by measuring how much the stars appear to move across the sky as the Earth orbits the Sun. This is called parallax, and in principle it is exactly the same as the effect you see when you hold up a finger at arm's length, and look at it while you close each of your eyes alternately. The image of the finger seems to jump across the background, because each of your eyes views it from a slightly different angle. The shorter the distance from your eyes to the finger, the bigger the effect; by measuring the angles, you could actually use triangulation to work out how long your arm is, if you were daft enough to try. This technique showed that even the nearest star to the Sun is 4.3 light years away. It takes light, travelling at 300,000 km a second (9.46 million million km a year), 4.3 years to get from the star nearest the Sun to us. And the Sun is close. Most stars are very much further away than that.

There are also ways of working out the average distance to all the stars in a cluster, moving together through space like a swarm of bees. If you watch for long enough – years in some cases – you can see how the stars move across the sky. Because the stars are in a cluster, moving more or less in the same direction through space, they seem to be converging on a point in the sky, the same way that parallel railway lines seem to converge on a point on the horizon. This tells you the angle the stars are moving at in three dimensions; once you know that, you can use fairly simple geometry to work out how far away they are from us in order to account for the sort of sideways motion we see. But this trick works only for relatively nearby clusters of stars, out to distances of about 100 to 150 light years.

Illustration 7. The principles of parallax and triangulation.
(Illustration copyright © 2003 Nicholas Halliday)

Parallax is the apparent displacement in the position or direction of a celestial body when it is viewed from different locations. It can be observed either simultaneously from two widely separated stations on Earth, or at intervals of six months from opposite sides of the Earth's orbit. By using triangulation, the resulting angles give the distance of the star or planet from the Earth. The greater a body's parallax, the closer it is to the observer.

Measuring the distance to the Moon

☆　☆　☆　☆　☆　☆

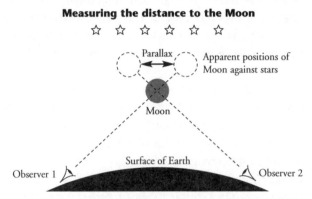

By measuring the angle between the two apparent positions of the Moon and the distance between the two observers, the Moon is calculated to be c. 384,400 km from Earth.

Measuring the distance to nearby stars

☆　☆　☆　☆　☆　☆

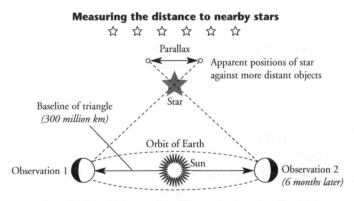

More distant bodies exhibit a much smaller parallax (greatly exaggerated here), requiring a considerably larger baseline for an accurate triangulation of the distance.

The good news is that if you know the distance to a whole cluster, which might contain hundreds of stars, you can work out how bright other stars with different properties (different colours and so on) really are. Then, if you see a similar star somewhere else, you can estimate how far away it is by guessing that it really is the same intrinsic luminosity as the stars in the original cluster and measuring its 'apparent brightness' – that is, how faint it looks.

These techniques are just as rough and ready as they sound. But there is one trick that opened up the whole business of measuring astronomical distances. This discovery was made by Henrietta Swan Leavitt at the Harvard Observatory in 1908 and was pinned down accurately as a reliable way to measure distances in 1912. There is a family of stars, called Cepheids, which vary their brightness in a regular way. Henrietta Swan Leavitt worked out that the time it takes for one of these stars to run through a cycle of brightening, fading and brightening again (its period) depends on the star's absolute brightness, averaging over its whole cycle. So if you see a Cepheid, you have only to measure its period to know how bright it really is. If you know its intrinsic brightness, you can work out how far away it looks from how faint it appears in the sky. There are just a few Cepheids in clusters close enough to have their distance measured, and these are used to calibrate the 'Cepheid distance scale'. By studying Cepheids in distant clusters it is possible to work out how far away those clusters are and to learn even more about how things like the colours of stars are related to their brightness.

Working on Mount Wilson from 1914 onwards, mainly with the 60-inch telescope, Harlow Shapley used this Cepheid method to work out distances to stars across the Milky Way, calculating in 1919 that we live in a huge disc

Illustration 8. The US astronomer Henrietta Swan Leavitt (1868–1921). After graduating from Radcliffe College in 1892, she joined the Harvard College Observatory in 1895. By 1912, she had discovered an accurate and reliable way to measure astronomical distances by using the 'period' of Cepheid stars. (Royal Astronomical Society)

of stars (the Milky Way) which is about 300,000 light years across, and that the Sun and its planets lie two thirds of the way to the edge of the disc. He was right about the shape of what is now called our Galaxy – it is like a fried egg, with a bulge in the middle surrounded by a thin disc – and he was right about our relative distance from the centre out into the galactic suburbs. But we now know that Harlow Shapley calculated the size of the Milky Way Galaxy incorrectly because he hadn't allowed for the effect of dust dimming the light from distant stars. A star may look faint because it is far away, or because some of its light is blocked by dust, or both. If it is faint because of dust, but you think it is faint because it is far away, you will calculate that it is further away than it really is. When dust is allowed for properly, the Milky Way turns out to be only 100,000 light years across – still big enough to contain around 200 billion stars each more or less like the Sun.

Harlow Shapley, together with many other astronomers in the second decade of the twentieth century, thought that this made up the entire Universe – everything that existed – and that the nebulae were either clouds of gas and dust inside the Milky Way, or little satellites orbiting the Milky Way in the same way that the Moon orbits the Earth. They also thought that the Milky Way had existed for ever, and that, although individual stars might grow old and die, they would be replaced by new stars, in much the same way that a forest can exist for much longer than the lifespan of an individual tree.

But other astronomers had different ideas. Some believed that the Milky Way was just one part of the

Universe, and that things like the Andromeda Nebula were other 'island universes', galaxies in their own right, so far away that even the light of hundreds of billions of stars put together added up to make only a faint patch of light in the sky. The only way to find out the truth was to develop telescopes and photographic techniques good enough to pick out individual stars in those nebulae (if they really were made up of myriads of individual stars). This is where Edwin Hubble and the 100-inch telescope won through, and where Harlow Shapley missed a chance of glory because of his conviction that nebulae were small nearby objects that did not contain stars.

Edwin Hubble had started his career at Mount Wilson by using the 100-inch telescope to develop the work he had begun during his years as a PhD student in Chicago. The many photographs of nebulae that he took using the Hooker telescope helped him to complete a classification system which arranged different kinds of nebulae in a sequence according to their shape. This is a bit like the way railway locomotives in the age of steam railways were classified in accordance with the size and number of their wheels. With hindsight we can see that the most important thing about all the observing that Edwin Hubble did for this project between 1920 and 1923 is that it made him thoroughly familiar with the 100-inch telescope. During that time he became a very skilled observer who could get the best out of the sometimes cranky and difficult telescope, coaxing it to produce superbly detailed photographs of nebulae.

The arguments about the nature of the nebulae still hadn't been resolved by the time that Edwin Hubble

Illustration 9. Edwin Hubble and Sir James Hopwood Jeans at the 100-inch Hooker telescope, with which Hubble became extremely familiar. (Henry Huntington Library, San Marino, California)

finished his classification project, so he decided to move on and attempt to solve it himself. Though not a man to lack self-confidence, even he didn't hold much hope of being able to pick out individual stars like the Sun in these nebulae, even using the powerful 100-inch telescope. At that time, though, he knew nothing about Milton Humason's observations that had been so short-sightedly dismissed by Harlow Shapley. But he thought of another way to measure the distances to nebulae.

In our own Galaxy, some stars occasionally flare up to as much as 100,000 times the brightness of our Sun over a few days, then fade away again over a few months.

They are called novae, because hundreds of years ago astronomers thought that they were literally new stars. Edwin Hubble figured that if there were novae in the Andromeda Nebula, and if he took enough pictures of the Nebula at different times, then he might be able to pick them out. All novae shine with very roughly the same brightness, so if he did detect novae in the Andromeda Nebula, he would be able to estimate the distance to the Nebula from how faint the novae looked. He began a new observing run with the 100-inch telescope in the autumn of 1923 and the results exceeded his wildest expectations. Within a few days, in October 1923, Hubble had found one definite nova in the outer parts of the Andromeda Nebula and two other spots of light that he suspected were novae. This encouraged him to dig out from the files a series of glass plates with photographs of the Nebula taken by different observers, including Milton Humason and Harlow Shapley, over the years.

Close examination of the plates showed that one of the 'novae' was actually a variable star, visible in some photographs but not in others over a span of years. Even better, it was a Cepheid, with a period of just under 31.5 days. Using the Cepheid distance scale, which was the standard way of measuring distances at the time, this meant that the Andromeda Nebula was a million light years away, far beyond the limits of the Milky Way. The modern calibration of the Cepheid distance scale, which is considerably different (partly because of the problem with dust that we mentioned), puts it at least twice as far away, more than two million light years from us – and this, we now know, is the closest large galaxy to our own.

Illustration 10. By examining a series of plates of the Andromeda Nebula, Hubble was led to conclude that the Nebula, in fact a galaxy, is at a far greater distance than had previously been calculated. (Henry Huntington Library, San Marino, California)

But the exact distance didn't matter. The important thing was that Edwin Hubble had proved that the Milky Way is just one galaxy, and that the Universe extends to unimaginable distances beyond it, with very many other galaxies each filled with a profusion of stars.

Over the next few months, Edwin Hubble found more novae and Cepheids in the Andromeda Nebula, and more Cepheids in several other nebulae, all pointing to the same conclusion. The evidence was discussed at a meeting of the American Astronomical Society held in January 1925 where everyone, including Harlow Shapley, agreed that there was now no doubt that the Universe was a much bigger place than it had previously been thought to be and that these nebulae were galaxies. In fact, some nebulae *are* clouds of gas inside the Milky Way, but to avoid confusion they are still called 'nebulae'. The nebulae beyond the Milky Way are now called galaxies.

3

FROM THE RED PLANET
TO REDSHIFTS

EDWIN Hubble had now achieved one of the greatest breakthroughs in the history of astronomy. But he knew that he had been able to measure distances only to a few of the nearest galaxies to our own Milky Way. He called his work 'a reconnaissance' and dreamed of being able to measure distant galaxies across the Universe, far beyond the range where Cepheids would be visible even to the powerful 100-inch telescope.

Hubble had read about the work of an astronomer called Vesto Slipher, who was working at the Lowell Observatory in Flagstaff, Arizona, and thought there was something in it that might be a clue worth following up. The Lowell Observatory had been set up in 1894 by a very wealthy businessman called Percival Lowell with one purpose only – to find evidence to prove that there is life on Mars. But as well as this one major obsession, Lowell was interested in how the planets formed in the first place. One fairly widespread idea among astronomers at the time was that many of the nebulae viewed through astronomers' telescopes might be swirling clouds of gas and dust that were in the process of settling down to form a planetary system like the Solar System. At the time, most people still thought that the Milky Way was the entire Universe and that these nebulae were clouds of stuff between the stars.

Vesto Slipher had joined the team of astronomers working at the Lowell Observatory in 1901, when he had been employed by Percival Lowell to investigate a particular kind of nebula with a spiral structure that looked like the pattern made by cream stirred into a cup of coffee. At the time, astronomers guessed that a spiral nebula was probably a spinning disc of matter that was settling down to become a planetary system. Slipher worked mainly with a new and, by the standards of the time, powerful telescope, which was a 24-inch refractor. Attached to it was a device for photographing the spectrum of light from a faint object. This combination of spectrum and photograph came to be called a spectrograph.

Spectrographs were a great astronomical breakthrough because the spectrum of light from an object showed what that object was made of and how fast it was moving. The basic spectrum of light is the colours of the rainbow: red, orange, yellow, green, blue, indigo and violet. But looked at under a microscope, it is possible to see that the spectrum is crossed by both bright and dark lines. These lines are produced at very specific places in the spectrum, at specific wavelengths, by the atoms of different elements – so sodium produces one pattern of lines, oxygen another and so on. Each pattern of lines looks like a barcode and is just as unique as a barcode. So by measuring the positions of the lines in the spectrum of light from a star – or anything else – it is possible to tell what it is made of. This is what fascinated Vesto Slipher. He wanted to find out what the nebulae were made of, and see if it matched the sort of stuff astronomers thought that the Solar System had been created out of. But he found something quite un-expected.

The characteristic lines in a spectrum are always produced at the same wavelengths for the same elements. But if the object emitting the light is moving towards us, the whole pattern of lines (the entire barcode) is shifted to shorter wavelengths, which means to the blue end of the spectrum – this is called a blueshift. And if the object giving out light is moving away from us, the pattern is shifted towards the red end of the spectrum – a redshift. The amount of the shift (either blue or red) tells you how fast the object is moving either towards or away from us.

To find out what he wanted to know, Vesto Slipher had to develop the technology to take spectrographs of spiral nebulae, which are very faint objects, by first practising on brighter things, like stars. To give you some idea of how he was working at the limit of what was technically possible at the time, to get any usable information from the observations of spiral nebulae photographic plates had to be exposed for up to 40 hours over several nights. Slipher's breakthrough came in 1912 when he obtained a series of spectrographs of the Andromeda Nebula. Eventually, these studies would show that such nebulae

The atomic elements present in a stellar body produce a characteristic banding in its light spectrum. Due to the arrangement of the electrons that surround the nucleus, atoms of a particular element emit or absorb light only at well-defined wavelengths. Light is emitted by atoms in the hot regions on the surface of a star, producing bright lines in the spectrum. Absorption occurs in cooler regions, further out from the surface or in the clouds of dust or gas surrounding a star, and produces darker lines.

A spectroscope can be used to split the incoming light from a star into its component colours, and the resulting photographic plates, or 'spectrographs', can show both what an object is made of and how fast it is moving:

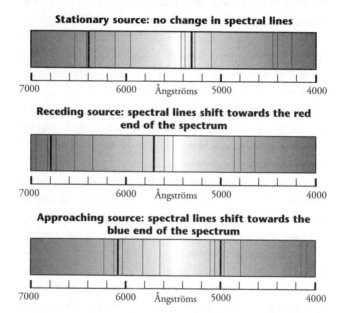

Stationary source: no change in spectral lines

7000   6000   Ångströms   5000   4000

Receding source: spectral lines shift towards the red end of the spectrum

7000   6000   Ångströms   5000   4000

Approaching source: spectral lines shift towards the blue end of the spectrum

7000   6000   Ångströms   5000   4000

The velocity and direction of travel of the star in relation to the observer determine the amount by which this banding shifts. When a radiating body is moving away from the observer, the waves emitted become 'stretched', the wavelength lengthens, and the spectral lines shift towards the red end of the spectrum. If the body is approaching, the wavelength is compressed, and the absorption lines shift towards the blue end of the spectrum. Redshift can be used to calculate an object's recession velocity.

Illustration 11. The shift of spectral lines. (Illustration copyright © 2003 Nicholas Halliday)

---

are made of stars, not just gas and dust – further evidence that they are galaxies. But what really stunned Slipher was the discovery that the entire spectrum from the Andromeda Nebula is blueshifted – so much so, that it means that the nebula is hurtling towards us at a speed of 300 km per second. At the time, this was the greatest speed ever measured for an astronomical object, and it shifted the focus of Vesto Slipher's work towards measuring more velocities for spiral nebulae.

As Slipher looked at more spirals, it became clear that the Andromeda Nebula was an exception. By 1914, he had measured velocities for a total of 15 of these objects, but only one other showed a blueshift. All the rest were redshifts, indicating that the objects are moving away from us, and two of them were moving away at speeds greater than 1,000 km per second. Vesto Slipher could measure these effects only in the spectra of the biggest and brightest spiral nebulae – even they were only just bright enough for their light to be analysed in this way. It seemed logical to guess that these must be the nearer spirals to us – but at the time nobody knew how near that meant. However, as Slipher slowly pushed the number of measured redshifts up, a hint of a pattern began to emerge. By 1925, just at the time that Edwin Hubble was reporting the first actual measurements of distances to what were now recognised as spiral galaxies, Vesto Slipher had pushed his score up to 39 redshifts and still just 2 blueshifts. From this evidence, it looked as if bigger, brighter galaxies had smaller redshifts. The best guess was that bigger, brighter galaxies are closer to us than faint small galaxies – rather, that they are all pretty much the same size but that some look big and bright because they are close and others look small and faint because they are far away.

Slipher could not probe any further out into the Universe to test this idea because he had reached the limit of what the 24-inch telescope and the technology available at that time could do. Edwin Hubble decided to pick up where Slipher had left off using the 100-inch

Illustration 12. A typical spiral galaxy showing characteristic spiral arms radiating from a denser nucleus. After joining the Lowell Observatory in 1901, Vesto Slipher developed the technology to take spectrographs of very faint spiral galaxies. (Chris Butler/Science Photo Library)

telescope on Mount Wilson. But this wasn't something that could be done overnight because, although the 100-inch reflector was more powerful than the Lowell Observatory's 24-inch refractor (even after allowing for the fact that, size for size, refractors are generally more powerful than reflectors), it had not been used for this kind of work before. It had been used for spectroscopy, but not on such faint objects, and it would take great skill and patience to develop its use for this kind of spectrographic study.

There was another problem. It was one thing to measure redshifts – it was difficult but possible – but quite another to measure distances. Edwin Hubble was the distance-measuring king as far as galaxies were concerned and, confident as ever, he wanted to keep it that way. He planned to use the 100-inch to measure distances to as many galaxies as possible by every technique possible. Then, if he had redshifts as well for the same galaxies, he could see if there really was a relationship between redshift and distance. If that worked out, all he would have to do in future would be to measure redshifts and use them to calculate distances. But he needed somebody else to do the tedious work of adapting the 100-inch for this kind of fine, spectrographic work and to carry out a programme of redshift measurements.

Milton Humason was the obvious choice for two reasons. First, he was the best observer on Mount Wilson and possibly the best in the world, and he was also patient, careful and accurate. But in Edwin Hubble's mind there must have been another advantage in choosing him as an assistant – because Milton, a high-school dropout and former mule driver and janitor, would undoubtedly be seen as just that, an assistant not an equal partner in the project. Always seeking to be the centre of attention, Edwin Hubble wanted the glory entirely for himself, and by and large he got what he wanted. But he could never have done it without Humason's help.

He only just got that vital help. Milton Humason agreed readily enough to test the capabilities of the 100-inch telescope for this kind of work. He did so by taking a spectrograph of a 'nebula' too faint for Slipher to have analysed with the 24-inch at Lowell Observatory. It took him two nights, high on Mount Wilson, observing through the 100-inch and produced a spectrum showing that the object was rushing away from Earth at about 3,000 km per second, nearly three times as fast as the largest recession velocity found by Vesto Slipher. Frankly, though, Milton Humason was not at all sure that it had been worth the effort.

This kind of observing was hard work. The telescope pointing up through the wide aperture in the dome of its building needed constant attention to keep the faint nebula in the centre of the field of vision, which meant getting cramped bending over the controls hour after hour. On Mount Wilson, the nights were cold even in summer, and in winter it was bitter and freezing. But the telescope dome could not be heated because that would set up convection currents in the air that would distort the image of the nebula. Most observing was done during the winter because that was when nights were longest, and it took two whole nights of painstaking (and literally painful) work to produce a single spectrum.

Illustration 13. An aerial view of Mount Wilson Observatory, showing the beauty and remoteness of its setting. The solar telescopes can be seen on the left, while the domes of the reflectors are on the right. (Courtesy of Carnegie Observatories, Carnegie Institution of Washington)

Edwin Hubble's project would need dozens of spectra like these, many of them for even fainter spirals, each requiring even more nights freezing at the controls of the telescope. Not surprisingly, Milton Humason initially turned the project down. But he was persuaded into having second thoughts by the promise that the Observatory would soon take delivery of an improved spectrograph that would make it possible to get pictures of faint spirals in a single night. He should have guessed what would happen. As soon as the new, improved spectrograph arrived, Edwin Hubble wanted to obtain spectra of even fainter galaxies that required Milton to carry out several nights observing, even with the new machine. But by then he had caught Edwin Hubble's enthusiasm for the project, and pushed both himself and the telescope to the limit in the search for higher redshifts – which, it soon became clear, really were associated with the fainter spirals.

But Milton Humason stayed cautious and steady. Exhibiting the extreme patience that had made him a good mule driver, after the first test of the ability of the system he spent months taking spectrographs of all the objects studied by Vesto Slipher, in order to be sure that the 100-inch, the new spectrograph and he himself were really getting things right, and also to double-check that Slipher hadn't made any mistakes. All the results matched. Only then was Milton Humason prepared to start work on a programme of obtaining 'new' redshifts.

4

THE EXPANDING UNIVERSE

WHILE Milton Humason was checking the redshifts that Vesto Slipher had found (which totalled 43, plus the two blueshifts, by the time Slipher had exhausted the capacity of the 24-inch), Edwin Hubble was measuring distances as accurately as he could to the same galaxies, plus the one extra spiral that Milton Humason had used to test the capability of the 100-inch. Edwin Hubble's work was just as tedious and back-breaking as the work Humason was doing – on the same telescope but on different nights – and unfortunately it soon became clear that only the very nearest spirals were close enough for their distances to be measured directly from Cepheids.

This posed problems which plagued attempts to measure distances across the Universe right up to the end of the twentieth century. The only way round it is to find some class of object that is brighter than a Cepheid, so it can be seen further away, but whose members all have the same brightness as each other. Then, the distances to the brighter objects can be tied in to the Cepheid distance scale by looking at galaxies like the Andromeda Nebula, where both Cepheids and the brighter objects are visible from Earth.

One of these so-called secondary distance indicators used by Edwin Hubble was the novae, which we have already mentioned. But he knew that novae do not really all have exactly the same brightness, so they could give only a rough guide to distance. Another was clusters of stars. Some star clusters contain millions of stars in a spherical ball. With all those stars shining together, the clusters can be picked out like individual jewels around some of the closer spirals. These became known as globular clusters; they come in different sizes, and don't all have the same brightness. But Hubble guessed that there might be a limit to how big a globular cluster could be, and to how bright it could be. By comparing the brightest globular cluster in one galaxy with the brightest globular cluster in another galaxy, he got a rough idea of how far away the second galaxy was compared with the first – twice as far, half as far or whatever.

Edwin Hubble tried every way he could think of to find the relative distances to more remote galaxies by comparing certain features in them to the same kind of features in galaxies like the Andromeda Nebula, where the distance was known directly, from Cepheids. If Hubble could say that a certain galaxy was, say, ten times further away than the Andromeda Nebula, that was all he needed to know – he didn't need to see Cepheids in the more distant galaxy to know its distance.

It wasn't perfect, but it was the best that could be done with the tools available. By 1928, Edwin Hubble had completed this first stage of the project, and together with Milton Humason had already begun to look deeper into space than Vesto Slipher. He wanted to wait a little longer before he presented his discoveries to the world, so that he would have a lot of new data and could make a big impression. But something happened that encouraged him to break cover sooner than he had intended.

Towards the end of 1928, the Swedish astronomer Knut Lundmark made a request to Walter Adams, who was then the Director of the Mount Wilson Observatory, asking to visit the mountain observatory so that he could use the 100-inch telescope to measure redshifts of nebulae. He even asked if Milton Humason would be able to act as his assistant. Knut Lundmark's request was politely turned down, but with this evidence that others were already hot on the same trail that he was following, Edwin Hubble decided that he ought to publish a scientific paper establishing his priority and revealing to the world his most important discovery.

The scientific paper announcing Edwin Hubble's observations and conclusions was published early in 1929. It was just six pages long and discussed only the first 46 galaxies studied. But in those six pages Hubble spelled out the fact that there really was a relationship between redshift and distance. What's more, it was the simplest kind of relationship – redshift is directly proportional to distance. So a galaxy twice as far away has twice as big a redshift, a galaxy three times as far away has three times as large a redshift, and so on. This became known as

Illustration 14. The Director of Mount Wilson Observatory, Walter Adams, with Edwin Hubble and the English mathematician, Sir James Hopwood Jeans, at the control panel of the 100-inch Hooker Telescope in 1931. (Courtesy of Carnegie Observatories, Carnegie Institution of Washington)

Hubble's Law. As we have seen, it might have been Slipher's Law, or Lundmark's Law. That is usually the way in science – new discoveries don't happen because of the unique ability of a single genius, but build on earlier work and take advantage of new technology. They happen because the time is ripe for them.

The time for discovering Hubble's Law was ripe in the 1920s because there were new telescopes and better spectrographs available, enabling astronomers to search further out into the Universe than anyone had looked before. Somebody was bound to make the discovery – Edwin Hubble was the right man, with a background in studying nebulae, working at the Mount Wilson Observatory, which was the right place at the right time (the 1920s). But the fact that the discovery was more or less inevitable doesn't mean that it wasn't hugely important. Edwin Hubble had finally established that to find out how far away a galaxy is, its redshift must be measured and divided by a number (now known as the Hubble Constant, or H) to give its distance.

At least, that was all you had to do in principle. The key problem now was to measure the value of the constant, H, accurately, by measuring distances to as many galaxies as possible, as far out into the Universe as possible, as accurately as possible, and measuring redshifts for the same galaxies. The next step down what was to prove a long road was taken in 1931, when Edwin Hubble and Milton Humason together published a longer and much more important paper which included data for another 50 galaxies, out to a redshift corresponding to a recession velocity above 20,000 km per second, and showed that the relationship (Hubble's Law) still held. On the basis of these measurements, they calculated the value of the key constant (which Hubble and Humason called K, but which we call H) to be 558, and the fastest moving galaxy in the sample was estimated to be a little over 100 million light years away from us.

We now know that for various reasons Edwin Hubble and Milton Humason's measurement of the value of H was almost exactly ten times too big. But whatever the exact value of H was, Hubble and Humason had discovered a profound truth about the Universe – that the galaxies are receding from one another, and that the velocities with which they recede are proportional to the distances between them. It isn't that we are at the centre of the Universe with everything rushing away from us. This simple redshift–distance relationship, Hubble's Law, that velocity is proportional to distance, is the only kind of relationship (apart from all the galaxies standing still) that looks the same whichever galaxy you are sitting on. Every galaxy in the entire Universe sees the rest of the Universe expanding away from it in line with Hubble's Law.

It didn't take long for some astronomers to point out that this meant that all the galaxies used to be closer together in the past, and that, if you went back far enough in time, they must all have been touching one another. Before that, all the stars must have been merged into one amorphous lump of stuff. The Universe must have had a beginning in the form of a hot fireball – a phenomenon that has now became known as the Big Bang.

5

THE BIG BANG

SEVERAL people contributed to the invention of the idea that we call the Big Bang, but the one who picked up directly from the observations made by Edwin Hubble and Milton Humason, and made the idea popular, was a Belgian priest, Georges Lemaître. He wasn't only a priest; in fact Georges Lemaître had the kind of genuinely exciting background that Edwin Hubble could only dream about.

Georges Lemaître had trained as a civil engineer at the University of Louvain, in Belgium, but he was just twenty in 1914 and volunteered to serve in the artillery when war broke out. His distinguished service as an artillery officer led to the award of the Belgian Croix de Guerre, and after the war he made a double change in career. First, he completed a doctorate in mathematics and physics at the University of Louvain, in 1920; then, he trained as a Roman Catholic priest, being ordained in 1923. He never practised as a parish priest, but became an important

51

member of what you might call the scientific civil service of the Church of Rome, eventually becoming President of the Pontifical Academy of Sciences.

It is easy to imagine, and many people have, that Lemaître's cosmological ideas, promoting the idea that the Universe was born (or created) at a definite moment in time, were coloured by his religious faith and the Christian image of 'the beginning' described in Genesis. But this would be doing him an injustice. The reason why Lemaître deserves pride of place in the story of the Big Bang is precisely because he did not base his scientific ideas on faith but constructed theories in accordance with observations of the real world.

The importance of having a link between theory and observations is highlighted by the story of what happened when Albert Einstein first tried to use the mathematics of his new General Theory of Relativity (more general than the older Special Theory, because it deals with gravity) to describe all of space and time (the Universe) in 1917. The equations said that the Universe must be expanding, or possibly contracting, but was definitely not standing still. At that time, nobody knew that the Universe was expanding, so Einstein thought that his equations were wrong. He fiddled with them, adding in an extra term to cancel out the expansion, and then moved on to other things.

In the 1920s, several mathematicians played with the equations Einstein had discovered which describe space and time and gravity. They looked at versions of the equations which described expansion, and at versions which describe contraction. But they did not do so because they thought the Universe really was like that; they did so because mathematicians like playing with equations. Georges Lemaître was different. He was also fascinated by the equations, and he calculated different ways in which space and time might behave. Above all, he wanted to know if the Universe really was like that.

After he had been ordained into the Roman Catholic church, Georges Lemaître was awarded a scholarship from the Belgian government which enabled him to travel to England, where he spent a year at the University of Cambridge, and then on to the United States, catching up on all the latest ideas in relativity theory and astronomy along the way. Once in the US, he worked for two years at MIT and with Harlow Shapley at Harvard Observatory, carrying out research for a PhD. Although his qualification from Louvain gave him the title 'Dr', in fact the work he had done there was about the same as a modern Master of Science degree, so it was well worth working hard for the American PhD. He also visited Edwin Hubble in California.

Georges Lemaître was actually present at the meeting where Edwin Hubble's discovery of the distance to the Andromeda Nebula was announced, and he kept in touch with developments on Mount Wilson after he returned to Belgium, where he was appointed Professor of Astrophysics at the University of Louvain in 1927. He immediately realised that if nebulae are galaxies, and the Universe extends far beyond the Milky Way, then the equations he had been intrigued by really could apply to the real Universe. Georges Lemaître was one of the first people to appreciate that on a truly cosmic scale galaxies like the Milky Way are merely 'test particles' which show how space is changing as time passes, like little chips of wood floating on a river and being swirled around by the currents.

The same year that he took up his post in Belgium, he published a paper suggesting that the Universe might really be expanding, and that the expansion might show up in measurements of the velocities of galaxies. But nobody paid any attention to this suggestion at the time, not least because it was published in an obscure Belgian journal that none of the astronomers working with the big telescopes bothered to read.

Things changed after Hubble published the discovery of the redshift–distance relationship (Hubble's Law) in 1929, and especially after Edwin Hubble and Milton Humason published their key paper in 1931. During his time in Cambridge, Georges Lemaître had met Arthur Eddington, the leading British astronomer of his time and widely recognised as the greatest authority on relativity theory after Einstein himself. In 1929, he sent a copy of his 1927 paper to Eddington, who realised its importance and arranged for it to be translated and published in a top English scientific journal. It actually appeared in print in English in 1931, after the Hubble and Humason paper, which helped to make sure that people took notice of it this time around.

In the 1930s, Georges Lemaître developed the idea that the Universe had a definite beginning in the form of a super-dense object that he called the 'primeval atom' (sometimes, the 'cosmic egg'). The name 'Big Bang' wasn't given to the idea until the 1940s when it was coined by the astronomer Fred Hoyle for a radio broadcast. But apart from that quibble, you can say that the Big Bang idea was born in 1931, and the father of the Big Bang was Georges Lemaître.

The first version of the Big Bang idea – Georges Lemaître's primeval atom – was directly based on the idea of winding the expansion of the Universe backwards and imagining (or rather, calculating) what things would have been like long ago. If you took all the stars in all the galaxies known to Lemaître and squashed them together to make a single lump of stuff with the same density as the nucleus of an atom, it would make a ball only about 30 times bigger across than the Sun is. This is absolutely amazing. Even though astronomers now think that there is a lot more matter in the visible Universe than could be seen from Earth in Georges Lemaître's day, even a ball ten times bigger across again (which means a thousand times bigger in terms of mass, because the volume goes as the cube of the radius) would only just stretch across the orbit of Mars around the Sun. And yet, our Solar System is tiny compared with the vastness of space.

The reason that matter can be packed up so small is that even inside atoms the amount of matter is tiny compared with the amount of empty space. This was only discovered at the beginning of the twentieth century, when the physicist Ernest Rutherford and his colleagues probed atoms by firing beams of tiny particles (called alpha particles, in streams called alpha rays) at them, and measuring how the alpha rays were deflected. They found that an atom is mostly empty space, with a cloud of electrons surrounding a tiny nucleus where almost all the matter is concentrated. In round numbers, the smallest atom is 10,000 times bigger across than the nucleus, and the largest atom is 100,000 times bigger across than the nucleus. Because of the cube rule for volume, this means that there is actually room inside one of those larger atoms for $100,000^3$ nuclei – you could fit one million billion nuclei inside a single atom.

There are only a couple of hundred billion stars in the entire Milky Way, and only a few hundred billion galaxies visible in the Universe. If you had 500 billion galaxies like the Milky Way, and you had one nucleus for every star in all of those galaxies, you would be able to fit them comfortably inside the volume of a single atom, with ten per cent of the space left over to give them some elbow room.

So Georges Lemaître's image of the birth of the Universe should really have been called the 'primeval nucleus', not the 'primeval atom'. In the 1930s, physicists were just beginning to investigate the way in which some unstable heavy nuclei split apart spontaneously. The proper name for this is nuclear fission, because it is the nucleus that breaks apart. But the popular name is 'splitting the atom', and it was in the same spirit that Lemaître called his primordial cosmic object a primordial atom rather than a primordial nucleus. Whatever the name, he was deliberately borrowing the idea of nuclear fission from the physicists. If such a huge primordial nucleus existed, he argued, it would be unstable and would break apart into smaller pieces that would fly away, creating the expansion of the Universe in a Big Bang (though he did not call it that), and, with the fragments breaking up more and more, eventually producing the atoms of the everyday world.

There's one thing wrong with the image this conjures up and one unanswered question that there was no way to answer at that time. The image of the primordial nucleus sitting in empty space, then exploding into fragments that fly away from each other through space, is wrong. The important thing about the mathematical description of the Universe based on Albert Einstein's equations is that it tells us that space itself expands and takes galaxies along for the ride. It is like those wood chips floating on the surface of a river. If the river is narrow the wood chips float along side by side, but when the river widens out as it enters a broad valley the wood chips may move apart from one another. They don't do so because they are moving through the water – they have no engines or oars, and for the sake of argument we can say that it is a still, windless day. They move apart because the amount of water between each chip increases – the river expands and carries them with it. And that is the way you should picture the expanding Universe. Space itself expands, and takes galaxies along for the ride. The amount of space between galaxies gets bigger, without the galaxies moving through space.

The unanswered question was where Georges Lemaître's primeval atom had come from. What was there before the Big Bang? It took more than 50 years for cosmologists to really get to grips with that one, and they are still argu-ing about the details. The arguments are inevitable, because it all happened about 13 billion years ago, and because, as cosmologists now think, it all has to do with quantum physics. Understanding quantum theory is difficult enough on its own, without bringing the birth of the Universe into the story. The main difference from Georges Lemaître's picture is that those quantum processes 'created' the seed of what became our Universe as a ball of pure energy, not matter, and that the matter came later, manufactured out of pure energy in line with the most famous equation in science, $E=mc^2$. But those details don't have anything to do with the story of how astronomers measured the distances to the galaxies and calibrated the size of the Universe. You don't need to know how the Universe was born in order to measure how big it is today, any more than you need to know when or where a person was born in order to measure their height.

6

A CONTINENT AMONG ISLANDS

EDWIN Hubble's discovery that the spiral nebulae are galaxies and his work with Milton Humason which showed that the Universe is expanding were just the beginning of efforts to measure distances across the Universe – to work out how far 'up' we can see. Edwin Hubble didn't worry too much about what the red-shift–distance relationship meant. It was natural to see this as evidence that the Universe is expanding, especially since this matches up perfectly with the equations of the General Theory of Relativity. In fact that is the way in which most astronomers, from Georges Lemaître onwards, had interpreted the relationship. But Edwin Hubble's big interest was distances, and what he really cared about was calibrating the relationship accurately, which means finding out the exact value of H, the Hubble Constant. Then he would be able to measure distances to any galaxy whose redshift could be measured, simply by dividing the measured redshift by the known value of H.

Edwin Hubble knew that even with a few dozen measurements of distances to galaxies he could have only a rough idea of the value of H. Apart from the difficulty of actually measuring things like the brightness of Cepheids or supernovae or globular clusters in distant, faint galaxies, there was another problem. The expansion of the Universe is not the only thing that can affect the light from distant objects. The way they move through space also shifts the lines in their spectra. This had been known since the 1840s when it was predicted by an Austrian physicist called Christian Doppler. Even today it is called the Doppler effect, in his honour.

Christian Doppler died as a result of a lung disease in 1853 when he was only 50. But in his short lifetime he had worked through four different professorships. In 1841, when he was 37, he was appointed Professor of Mathematics at the Technical Academy in Prague, moving on in 1847 to become a professor at the Mining Academy in Chemnitz, in Germany. In 1849 he moved back to Austria to take up the job of Professor of Geometry at the Vienna Technical University before becoming Professor of Experimental Physics at the university there in 1850. Clearly a life that left little time for non-academic adventures.

The Doppler effect is very familiar to most people through its effect on sound waves. As a fast-moving fire engine travels towards you and sounds its siren, you hear a higher note than when the same vehicle has gone past and is speeding away from you. This is because the sound waves from the siren get squashed up as the vehicle approaches you, but are stretched out when it is moving away. There were no fire sirens and very few fast-moving vehicles in Doppler's day, so, after he predicted this effect on the basis of his calculations about the behaviour of sound waves (made in 1842), he tested the prediction using a steam engine pulling an open truck along a straight railway line running across flat land in Holland. In the truck, there were skilled trumpeters who played a constant note as loudly as they could while the train went by. On the trackside, there were musicians who had perfect pitch and could tell you the precise note they heard. The difference in the notes they heard when the train was coming towards them and when it was going away from them exactly matched the effect calculated by Christian Doppler.

The same effect also applies to light, and for the same reason. If an object is moving towards you, the light waves it gives out are squashed up by the motion – a blueshift. If the object is moving away from you, the light waves are stretched out – a redshift. But in order for the effect to be noticeable, the object has to be moving at a good fraction of the speed of light, which is 300,000 km per second. So we don't notice it in everyday life, only in astronomy.

It is important to realise that the cosmological redshift, as the effect discovered by Edwin Hubble is usually called, is *not* strictly a Doppler effect, because it is not caused by galaxies, or anything else, moving through space. It is caused by space itself stretching, and stretching the light passing through it. So when a galaxy moves through space, either towards us or away from us, it will produce an extra effect, a true Doppler effect, which is added on to the cosmological redshift – or maybe subtracted from it.

That is why the first spiral nebula studied by Vesto Slipher, the Andromeda Nebula, actually shows a blueshift in its light. The Andromeda Galaxy, as it is now called, is the nearest large spiral to us, and since the cosmological redshift is proportional to distance, the galaxy's cosmological redshift is tiny. Because the Andromeda Galaxy is moving towards us through space, the blueshift this produces by the Doppler effect completely overwhelms the cosmological redshift. It is as if the Andromeda Galaxy is represented by a speedboat ploughing its way upstream, against the current. The river is trying to carry it off in one direction, but the boat is moving through the water in the other direction faster than the water can carry it downstream. But although the cosmological redshift is not a Doppler effect, astronomers often refer to it in terms of velocity, as if it were a Doppler effect. They will say that a galaxy 'has a redshift of 1,000 km per second', or whatever, meaning that the shift in its spectrum is the same size that would be produced by a Doppler effect if the galaxy were moving through space at that speed.

Of course, the Andromeda Galaxy doesn't have an engine attached to send it rocketing towards us. It is moving towards us because both the Andromeda Galaxy and the Milky Way, and a few other small galaxies, are part of a cluster, or group, of galaxies (called the Local Group) that are held together by gravity. These galaxies are all moving around under their mutual gravitational influences, being tugged this way and that by the gravity of the other galaxies. It just happens that the Andromeda Galaxy is travelling in our direction through space at a speed of about 300 km per second. But there is no need to worry about the possibility of a collision. The Andromeda Galaxy is more than two million light years away, and any impact won't happen for a very long time indeed.

But this does give you some idea of the typical sort of speeds galaxies move at within clusters like the Local Group. And most galaxies do indeed move in clusters, many of them much, much bigger than the Local Group. The redshifts and blueshifts produced by their movement, at speeds of several hundred kilometres per second, make it very difficult to unravel the cosmological redshifts from measurements of their spectra, unless you look so far out into the Universe that the cosmological redshift completely overwhelms the Doppler effect. But, yet again, that means looking at more distant, and more faint, galaxies, which makes the whole thing harder.

If a cluster of galaxies were at just the right distance for their cosmological redshift to be 300 km per second, then any galaxy in the cluster moving towards us at the same speed as the Andromeda Galaxy does might show up as having either zero redshift, if it is moving towards us through space, or a redshift equivalent to 600 km per second, if it is moving exactly away from us through space. If the galaxy is moving at an angle to our line of sight, only part of the Doppler effect shows up from Earth. And it wouldn't be quite so bad if all galaxies did move at the same speed, whether it was 300 km per second or any other speed. In fact, there is a whole range of speeds – one galaxy in a cluster may be moving towards us at 100 km per second, another may be moving away at 426 km per second, and so on – and we simply don't know what those speeds are.

But if a cluster of galaxies were at just the right distance for the cosmological redshift to be 3,000 km per second, and if the average speed of the galaxies in the cluster was the same as the speed that we see the Andromeda Galaxy moving relative to us, then the biggest effect they could have on our measurement of the Hubble Constant would be 10 per cent, since 300 is just 10 per cent of 3,000. From the time of Edwin Hubble to the end of the twentieth century, this has been the hope of cosmologists – to measure H and the distance scale of the Universe to an accuracy of 10 per cent.

Edwin Hubble thought he had done it in the early 1930s. But there was something odd about his early measurements of the distance scale which seems to have worried only one person at the time. That person was Arthur Eddington, who made huge contributions to the development of twentieth-century astronomy and practically invented the branch of science known as astrophysics, the physics behind how stars work. He had been born in 1882 and was brought up in a Quaker family. An outstanding scholar, he progressed smoothly up the academic ladder to become Plumian Professor of Astronomy and Experimental Philosophy in the University of Cambridge in 1912, and in 1914 he took on the additional duties of Director of the Cambridge Observatories. Being already in his thirties, and holding such important posts, you might think he was safe from getting personally involved in World War I. For a time, that was the case. Indeed, Arthur Eddington was able to maintain scientific contact, of a sort, with colleagues in Germany, even though Britain and Germany were at war. In 1916, when the scientific papers describing Albert Einstein's General Theory of Relativity were published, Einstein, who was then working in Berlin, sent copies of them to his friend Willem de Sitter in neutral Holland. De Sitter, in turn, sent copies of the great work on to Arthur Eddington in Cambridge, which is how Eddington became an expert in relativity theory almost as soon as Albert Einstein had come up with it. This explains why Einstein's theory was widely known about in England – thanks to Eddington spreading the word about it – before the end of the war.

Albert Einstein's theory predicted that light from distant stars, grazing past the edge of the Sun, would be bent by a certain amount. You can't usually see this, because the starlight is blotted out by sunlight. But during a total eclipse of the Sun, when the Moon sits right in front of the Sun, the sunlight is blocked out, the sky goes dark, and you can see those all-important stars. Arthur Eddington realised that in May 1919 there would be an ideal opportunity to test Einstein's theory, during an eclipse that would be visible from Brazil and from the island of Principe, off the western coast of Africa.

In 1917, at Arthur Eddington's suggestion and with the enthusiastic support of the Astronomer Royal Sir Frank Dyson, provisional plans were made to send two expeditions by sea to observe the eclipse in 1919. But with the war in Europe going badly, conscription was introduced in Britain, and all able-bodied men, even eminent professors in their thirties, were eligible to be called up for active military service. Many scientists were appalled at the prospect of sending the best brains of their generation to die in the trenches, and a campaign was organised to gain exemption for people like Arthur Eddington, who would be much more useful to their country doing their normal work. After a great deal of argument and discussion, the Home Office finally agreed to give exemption to some people working in universities and observatories. They wrote to Arthur Eddington to tell him that if he signed and returned a copy of the letter to acknowledge its receipt, he would not be expected to enlist. He did so, but added a footnote saying that as he was a Quaker and a pacifist, he would not fight anyway, even if he was ordered to do so by the government.

This upset some people at the Home Office. In 1917, conscientious objectors were treated harshly, and if they refused to join the armed forces they were sent to labour camps, and some were even sent to prison just for refusing to fight in a war. So if Arthur Eddington wanted to insist on being treated as a conscientious objector, as many of his Quaker friends had been, then the Home Office would take him at his word.

Fazed and irritated by what had happened, Arthur Eddington's friends rallied round again. After some smooth talking, Sir Frank Dyson, who held an appointment given by the Crown (Astronomer Royal), managed to achieve a compromise that saved face for the government. It was decided that Arthur Eddington was to be deferred from conscription into the Armed Forces, but only on condition that as a public service he *must* lead an eclipse expedition to test Einstein's theory, provided the war ended in time.

Illustration 15. Arthur Stanley Eddington (1882–1944), seen by many as the father of modern theoretical astrophysics, in 1927. His expedition to Principe in 1919 helped prove Einstein's General Theory of Relativity, on which he later became an expert. (Segrè Collection/American Institute of Physics/Science Photo Library)

It was thanks to this masterpiece of diplomacy that Arthur Eddington did indeed sail to Principe in the spring of 1919. The Armistice, bringing an end to the fighting, had been signed on 11 November 1918, though formally the war did not end until the Versailles Peace Treaty came into force, on 10 January 1920. This led to the curious fact that a German theory had been publicly proved right by a British expedition, while the two countries were still technically at war.

By the early 1930s, Arthur Eddington was the 'grand old man' of European astronomy (although he wasn't 50 until 1932), and like most scientists of that age he was no longer making important contributions to original research. But he did have a lot of experience to draw on, and he was nobody's fool. He was reluctant, for perfectly good reasons, to take the news from the other side of the Atlantic entirely at face value.

Because you work out the distance to a galaxy by dividing its redshift by the Hubble Constant H, the bigger the value of H is, the shorter the distance to that galaxy is. Edwin Hubble worked things out the other way round. He measured distances to galaxies, and Milton Humason measured the redshifts of the same galaxies, then Hubble calculated H. From that point of view, the closer a galaxy with a particular redshift is to us, the bigger H must be. Either way, it is an inverse relationship – big values of H go hand in hand with a short distance scale for the Universe.

The value of H that Edwin Hubble came up with in 1931 was 558. This corresponds to a short distance scale, with one surprising feature that Arthur Eddington drew attention to in a book, *The Expanding Universe*, published in 1933.

Because the distances that Edwin Hubble had been measuring to spiral galaxies were relatively short, that meant that the galaxies were not very big, compared with our own Milky Way Galaxy. When astronomers look at or photograph a spiral galaxy, the size of the image they get, before it is magnified, depends both on how big the galaxy really is and how far away it is. For example, the Moon and the Sun both look exactly the same size on the sky, a remarkable coincidence which is dramatically clear during a solar eclipse. If you hold your arm out straight in front of you, and bring your finger and thumb together to leave a gap just big enough for the Moon to fit into, the gap will be about the size of a pea. So a pea at arm's length looks the same size as the Moon, which is actually 3,476 km in diameter and 384,400 km away. And it is also the same angular size as the Sun, which is actually 1.4 million km in diameter, and is 149.6 million km away. If you see something on the sky that looks the size of a pea and you know how far away it is, then you can work out how big it really is. The same goes for any-thing – if you know angular size and distance, then you can work out its actual (linear) size.

When people like Arthur Eddington did this in the 1930s, using Hubble's distances to spiral galaxies and their measured angular diameters, the spirals all came out as having much smaller diameters than our own Milky Way, which can be measured from the inside using Cepheids and other distance indicators. This didn't seem to bother Edwin Hubble, who may have rather liked the idea of living in the biggest galaxy in the Universe. But it certainly bothered Arthur Eddington, who wrote in his book that the implication was that the Milky Way is like a continent among islands, and went on to say, 'Frankly, I do not believe it; it would be too much of a coincidence. I think that … ultimately we shall find that there are many galaxies of a size equal to and surpassing our own.'[1]

How could that be? One possibility is that there are other 'continents' in space, extra-large galaxies like the Milky Way, each surrounded by their own 'islands', but that even with the 100-inch Hooker telescope Edwin Hubble and Milton Humason had not been able to see far enough out into the Universe to find them. Another possibility is that Hubble's distance measurements were wrong, and that all the spirals he had studied were much further away than he thought – but they would have to be nearly ten times further away, reducing the value of the Hubble Constant to about 55 or 60, in order to imply that other spirals really are about the same size as the Milky Way. Nobody suggested such a drastic increase in the cosmic distance scale in the 1930s. But everybody knew that any questions about the reliability of Edwin Hubble's pioneering work could best be answered with even bigger telescopes, able to photograph even fainter objects further out into the Universe and take their spectra to measure the redshifts. Happily for the astronomers, thanks to the obsessive commitment of George Ellery Hale to building ever bigger and better

---

[1] Eddington, Arthur, *The Expanding Universe*, 1933, p. 5.

telescopes, a project to build just such an instrument, with a main mirror 200 inches in diameter, was already under way at Mount Palomar, a little to the south of Mount Wilson.

7

DOUBLING THE DISTANCE SCALE

GEORGE Ellery Hale really was obsessed by telescopes. He had started out by building the world's biggest refractor at the Yerkes Observatory. Then he moved on to Mount Wilson where he was the driving force behind the creation of a whole new observatory with first the 60-inch reflector then the 100-inch, each one in its time the best telescope in the world. By 1923, it might have seemed to some onlookers that George Ellery Hale had done enough and could sit back and view his achievements with satisfaction. He was 55 years old, and he had seen the Mount Wilson Observatory completed. The task had been a huge strain, both because of the difficulties of building an observatory on a mountain-top in those days, and because of the effort Hale had to put into fundraising and bureaucracy to keep the project alive. If you'd known him, you would have been surprised that he had managed so much – he had always suffered from depression and severe headaches, and over the years he had

Illustration 16. George Ellery Hale became Professor of Astronomy at the University of Chicago in 1892 at the age of 24. His obsession with the construction of better telescopes eventually led to several nervous breakdowns, but not before he had set in motion plans to construct a 200-inch reflector at Mount Palomar. (Royal Astronomical Society)

had to have operations for acute appendicitis, a diseased gall bladder and a kidney infection. He had also had three minor nervous breakdowns. It is hardly surprising that, after all that, he suffered a more severe nervous breakdown in 1922 and had to resign from his post as Director of the Mount Wilson Observatory on medical grounds the following year.

Hale's doctors advised him that he needed a quiet, restful life, and that he wouldn't ever be able to take up his old post again. He retired to his home in Pasadena, just below Mount Wilson, where his idea of a quiet life involved building a small observatory for his private use and inventing a better kind of spectroscope, which he used to study the Sun. Soon, even this was not enough to satisfy his restless ambition and he was off on his fundraising again, this time trying to get money to build an observatory in the southern hemisphere. This would have been a really valuable contribution to astronomy because large parts of the sky can be seen only from the southern hemisphere, and by the mid-1920s most of the astronomical observatories in the world were in the northern hemisphere. But this time even George Ellery Hale's powers of persuasion were not enough to get the project up and running, and he suffered another nervous breakdown as it collapsed.

For some time – ever since the 100-inch had got the go-ahead – Hale had been dreaming of building an even bigger telescope, with a main mirror at least 200 inches in diameter. As he recovered from his latest illness, he decided to try to make this last dream a reality, and enlisted the help of Francis Pease, one of the technical experts on Mount Wilson, to put the dream into practice. Francis Pease was so fired up by Hale's enthusiasm that he drew up plans for a 300-inch reflector, nearly ten times as powerful as the 100-inch (not just three times as powerful, because the area of the mirror goes as the square of the diameter).

Armed with Francis Pease's drawings, and with his powers of persuasion obviously fully restored, in 1928 George Ellery Hale took the scheme to the Rockefeller Foundation who agreed in principle to fund the new telescope. The project was scaled down to a 200-inch (roughly 5-metre) diameter main mirror, because even Hale had to admit, when he looked into the practicalities, that the technology to build a bigger telescope simply did not exist at the time. In May 1928 the Rockefeller Foundation formally offered to pay $6 million for the project. This was a huge amount of money at 1928 prices – enough to build a small town.

Delayed by technical problems, the Depression and World War II, the new telescope built on Mount Palomar did not become fully operational until 1948. That was ten years after George Ellery Hale had died, a few months short of what would have been his seventieth birthday. In his memory, the 200-inch was formally named the Hale Telescope.

Without doubt, the Hale Telescope was decades ahead of its time, and speeded up the advance of our understanding of the Universe by 20 or 30 years. It was only in the 1980s that other telescopes began to surpass its ability to probe the Universe, and even today it is still a major piece of kit used by astronomers working at the cutting edge of research. All thanks to the obsession of one man. And the power of the new telescope was shown to the world at large by some of the very first observations made with it, in the late 1940s.

The person who made those observations, directly benefiting from Hale's vision and the generosity of the Rockefeller Foundation, was Walter Baade, a German by birth, who had joined the staff at Mount Wilson in 1931. Astronomy is probably the most international of all the sciences, and usually where somebody is born doesn't have much to do with their work. But in this case Baade's nationality is important to the story, because it was a major factor in explaining why he had almost unlimited use of the best telescopes in the world at a time when most astronomers were preoccupied with other matters.

Walter Baade was a first-class astronomer, but not very good at organising his private life. He always intended to become an American citizen, but somehow never got round to it. It took him until 1939 to start getting the necessary paperwork in order, and he promptly lost those papers during a house move. So when the United States entered the war against both Germany and Japan after the Japanese attack on Pearl Harbor in December 1941, Walter Baade automatically became what was known as an 'enemy alien' – a citizen of Germany, which was at war with the US, but living in California. What made things worse, in the eyes of the authorities, was that Walter Baade's brother was a member of the Nazi party and a U-boat captain.

Most of the top scientists in the US, including the astronomers on Mount Wilson, were quickly recruited into war work. Edwin Hubble himself left for Maryland in the summer of 1942, where he worked in ballistics. But

Illustration 17. The 200-inch Hale reflector at the Palomar Observatory of the California Institute of Technology. It was dedicated to the memory of George Ellery Hale, whose leadership and vision led to its creation. (Royal Astronomical Society)

there was no way an enemy alien could be trusted with that kind of work. Quite the reverse. In the aftermath of the attack on Pearl Harbor, the entire west coast of the US seemed vulnerable, and Los Angeles was considered a potential target and possible war zone. Like the few other enemy aliens who were allowed to remain in the area, Walter Baade had to stay in his home at night and was allowed to travel only a few miles away in the daytime. But as the likelihood of a sea-borne attack on Los Angeles receded and things settled down into a wartime routine, after a few months the authorities reconsidered these restrictions. The fact that Walter Baade had started the process of applying for US citizenship back in 1939 (even if he had lost the paperwork) was taken as a sign of good faith, and he was allowed to go back to the observatory on Mount Wilson at night.

Walter Baade must have thought he was in astronomical heaven. He was able to use the 100-inch telescope almost any time that he wanted to, but that was only part of the story. In addition, a new kind of photographic plate, more sensitive than anything used before, had just become available and, as a bonus, the lights of the nearby city were blacked out because of the war. The result was that he was able to photograph stars fainter than anything that had ever been photographed from Earth before. And this led to an amazing discovery.

He found that there are two kinds of star – what he called two 'populations'. In 1944, Walter Baade announced that stars like the Sun occur in the spiral arms and disc of a galaxy like the Milky Way or the Andromeda Galaxy, and called them Population I. Population I stars tend to be hot, relatively young, and contain a lot of interesting elements like carbon and oxygen, revealed by their spectra. The central part of a galaxy like the Milky Way or the Andromeda Galaxy is made up of lots of older, redder stars that contain hardly anything except hydrogen and helium. They are called Population II. Population II stars also form the globular clusters that surround the entire galaxy.

The explanation for the two kinds of star is that Population II stars came first, formed out of the raw material from which the galaxy was made, and that Population I came later, formed out of material which had been partly cooked inside the first stars and then thrown out into space in stellar explosions (novae, and even bigger stellar explosions called supernovae). More importantly, Walter Baade also found that there are two different kinds of Cepheid-like variable star, one for each of these populations. The original Cepheids became known as classical Cepheids, and are all Population I stars, but there is another family in Population II. These are now called W Virginis stars, and they are fainter than the classical Cepheids. All the distance measurements to spiral galaxies were based on the key first step that Hubble had taken in the winter of 1923/4, measuring the distance to the Andromeda Galaxy using Cepheids. But what if the stars he had used in his calibration weren't all Cepheids, but a mixture of Cepheids and W Virginis stars?

The only way to find out would be to identify another kind of star in the Andromeda Galaxy. Walter Baade knew just the stars to look for. They are called RR Lyrae stars and are extremely good distance indicators because, although they vary in a regular fashion like Cepheids, they all have almost the same intrinsic brightness. But they are much fainter than Cepheids (or even W Virginis stars), and his only hope was that they might just be detectable in the Andromeda Galaxy using the new 200-inch Hale telescope.

It was a long project, and Walter Baade did other things along the way. The 200-inch did not become operational until 1948, and immediately there were many demands on its time from the astronomers back from the war and a new generation of observers. Walter Baade got his ration of time on the 'Big Eye', but even pushing the telescope to its limit, he could not identify RR Lyrae stars in the Andromeda Galaxy. He checked and double-checked his observations, but was forced to conclude that the Andromeda Galaxy was so far away that RR Lyrae stars could not be detected there, even with the 200-inch. How far was that? There was one big clue. In the course of his work with the 100-inch, Walter Baade had confirmed a relationship between the brightest Population II stars and RR Lyrae stars. In globular clusters in our Galaxy, the very brightest Population II stars (known as red giants) are all about the same brightness (which makes sense, since you might guess there is some limit to how bright a star can be), and this brightness could be compared with the average brightness of RR Lyrae stars. With the 200-inch, Baade could just detect the equivalent red giants in the Andromeda Galaxy. So he knew that the RR Lyrae stars he *couldn't* see were a certain amount fainter than these stars, and he knew how far away they must be to be so faint by comparison with RR Lyrae stars in our Galaxy. The distance was twice as far as Edwin Hubble had calculated in the 1920s – Hubble really had been fooled because of a confusion over classical Cepheids and W Virginis stars.

The discovery was announced at a meeting in Rome in 1952, more than 25 years after Edwin Hubble first measured the distance to the Andromeda Galaxy. It meant that the Hubble Constant had to be halved (to about 250), and the distance scale of the Universe had to be doubled – every galaxy distance measured before 1952 had to be multiplied by two. As the newspaper headlines of the day put it, the known 'size of the Universe' had doubled overnight.

But this was only a first step. All that had actually been pinned down for certain was the distance to the Andromeda Galaxy, a little over two million light years. And the Andromeda Galaxy is much too close to be used to calibrate Hubble's Law directly – so close that it is actually moving towards us, not receding. The telescope that could do the job of calibrating the distance scale properly was at last up and running, and had proved its worth. But by 1952, Edwin Hubble and Milton Humason were in their sixties, and even Walter Baade was 59. The time had come to pass the torch on to a new generation of astronomers. And the man who was to become Edwin Hubble's scientific heir was already at Mount Palomar.

8

HUBBLE'S HEIR

THE man who picked up the cosmological quest where Edwin Hubble left off was Allan Sandage, who wasn't even born until June 1926, a full 18 months after Hubble's discovery of Cepheids in the Andromeda Galaxy had been announced. Allan Sandage was hooked on astronomy by the age of nine, when he looked at the stars through a friend's telescope and persuaded his father to buy one just like it. He looked through that telescope at stars night after night. Edwin Hubble's book *The Realm of the Nebulae* was published in 1936 and keenly read by the young Sandage a couple of years later. He also read the popular books about astronomy and relativity theory written by Arthur Eddington (including *The Expanding Universe*), and was well aware that he was living through a revolution in our understanding of the Universe.

Eager to become an astronomer himself and take part in the revolution, Allan Sandage enrolled as a physics student at Miami University in Ohio, where his father was a teacher. He completed two years of his course before being drafted into the Navy, where he spent 18 months as a technician working on radar and radio equipment before being discharged in 1946, at the end of the war. By then, his parents had moved to the University of Illinois, and Allan joined them in order to live at home while finishing his education. He still studied physics, but volunteered to help out with a nationwide programme of star counting. This involved photographing a patch of the night sky and then examining the photographic plates to calibrate the brightnesses of different stars.

Allan Sandage had dreamed of working at Mount Wilson ever since 1941, when his father (who spent a summer teaching at Berkeley, near San Francisco) had taken him to the mountain to see the telescope that Edwin Hubble and Milton Humason had used to discover the expansion of the Universe. In 1948 when Allan Sandage completed his degree in Illinois, he wanted to work on Mount Wilson but had no idea how he could get a job there. So he applied to do a PhD in physics at Caltech, in Pasadena, because it was the closest educational establishment to Mount Wilson. The application paid off. Unknown to Sandage, the Caltech authorities had decided to start a new PhD in astronomy, admitting just five students that year, to begin training new observers and theorists to make best use of the telescopes on Mount Wilson and Mount Palomar and the information they expected to flow down from the mountains. The decision was so recent that the opportunity had not even been advertised.

With his experience as an observer, Allan Sandage was

Illustration 18. Allan Rex Sandage, photographed here in 1954, made numerous observations to attempt to fix the value of the Hubble Constant. He eventually showed that a major modification was necessary, as some of the 'stars' Hubble had identified in distant galaxies were actually HII regions. (© Estate of Francis Bello/Science Photo Library)

a natural for the new PhD programme, so when he was asked if he wanted to be one of the five students, he had no hesitation in accepting.

He arrived at Caltech to start training to become a professional astronomer the year that the 200-inch telescope became operational. At that time, Walter Baade was king of the mountains, as far as observing was concerned, and Edwin Hubble and Milton Humason were nearing the end of their careers. But Edwin Hubble was still very active in research, and Allan Sandage's experience at calibrating stars back in Illinois made him the obvious candidate when Hubble needed help for another of his pet projects, counting the number of galaxies with different brightnesses in different parts of the sky to get a better grasp of the geography of the Universe.

Allan Sandage was summoned to see Edwin Hubble in the spring of 1949, which must have been incredibly intimidating for him. The meeting went well, and Edwin Hubble set Allan to work. This work didn't involve actually using the big telescopes but sifting through stacks of photographs and picking out the objects of interest. But, in July, Hubble suffered a heart attack and was ordered to give up observing for at least a year. Although Allan Sandage didn't have the skill and knowledge to complete the project on his own, he was still keen to develop his observing skills. So he readily accepted when he was offered work helping Walter Baade with a project investigating the nature of globular clusters, one of the things that kept Walter Baade busy alongside his work on the distance scale. Allan Sandage and a fellow student, Halton Arp (usually known as 'Chip'), were taught how to observe properly by Walter Baade himself, using the 60-inch telescope on Mount Wilson. A little over 30 years before, it had been the best telescope in the world; now, although still being used for cutting edge research, even students had access to it. Allan Sandage's dream of being a great astronomer had started to come true.

His skill as an observer, quickly learning everything Walter Baade could teach him, did not go unnoticed. The following year, he received the ultimate accolade. Edwin Hubble, still too ill to go back up to the mountain observatory on Mount Wilson, decided he had to have an assistant to do the observing for him. So he sent Allan Sandage up Mount Palomar, with Milton Humason, to learn how to operate the 200-inch telescope on his behalf. At the age of 24, without having yet formally completed his PhD, Allan Sandage became one of the main observers with the Big Eye, acting as Edwin Hubble's eyes and hands.

All of this work was in addition to the project that Allan Sandage was working on for his PhD, which involved the nature of stars and their evolution – how they change as they age. But Edwin Hubble gradually returned to observing, so Sandage had more time to complete his project. In 1952, just when Allan Sandage was finishing this work and about to receive his degree, he was offered a permanent job on Mount Wilson. After taking a year out to visit colleagues at Princeton University to develop some of the ideas in his PhD thesis, he returned to California to take up the appointment in the autumn of 1953. He planned to carry on studying the way stars work, which was his great interest. But in September, soon after he arrived back in California, news came that Edwin Hubble had suffered a stroke and died.

It soon became clear that there was only one person competent to take over Edwin Hubble's programme of observing the Universe. It wasn't what Allan Sandage wanted to do – he wanted to study stellar evolution. But it was what he *had* to do. As he later told the science historian Alan Lightman:

I felt a tremendous responsibility to carry on with the distance-scale work. He had started that, and I was the observer and I knew every step of the process that he had laid out. It was clear that to exploit Walter Baade's discovery of the distance-scale error, it was going to take 15 or 20 years, and I knew at the time it was going to take that long. So I said to myself, 'This is what I have to do.' If it wasn't me, it wasn't going to get done at that period of time. There was no other telescope; there were only 12 people using it, and none of them had been involved with this project. So I had to do it as a matter of responsibility.[2]

Allan Sandage had greatness thrust upon him.

In fact, Allan Sandage did continue his studies of globular clusters and the way stars age, alongside his monumental work on the distance scale. And there was a plus to being Hubble's scientific heir. For Allan Sandage this came in the form of Edwin Hubble's allocation of time on the big telescopes, which included 35 nights a year on the 200-inch alone. Not bad for a 27-year-old astronomer who had completed his PhD little more than a year earlier.

---

[2] Lightman, Alan, and Brawer, Roberta, *Origins: The Lives and Worlds of Modern Cosmologists*, Cambridge, MA: Harvard University Press, 1990, p. 74.

But he was facing a monumental task. Edwin Hubble had had the right idea, trying to identify bright objects in nearby nebulae such as the Andromeda Galaxy, and measuring their brightness so that he could use them as 'standard candles'. The standard candles were calibrated against the known (now corrected) Cepheid and RR Lyrae distance to the Andromeda Galaxy, and then the faintness of the equivalent standard candles in more distant galaxies told you how far away they were. The problem was that, although Hubble had had the right idea, the tools he had were inadequate. Even the 100-inch wasn't powerful enough to see far enough into the Universe to calibrate the Hubble Constant properly. Allan Sandage had to spend years working with the 200-inch just gathering data – photographing galaxies and taking spectra – before he had enough information to build up a coherent picture.

The search for the cosmic distance scale got more and more vague the further out into the Universe astronomers looked. Beyond the range where Cepheids could be seen, Edwin Hubble had used what he thought to be individual bright stars as his standard candles. Further out still, even a globular cluster had to be regarded as a single candle. And at the limit of what he could survey with the 100-inch, he made the rough-and-ready guess that if you consider only galaxies that look alike, whole galaxies might have roughly the same brightness as each other, so that their overall faintness could be related to distance.

With the evidence coming in from the 200-inch, Allan Sandage found that many of these steps out into the Universe were imperfect. For example, in our Galaxy there are huge clouds of hot gas, with several stars embedded inside them, that are known as HII regions. The 200-inch revealed that what Edwin Hubble had thought to be bright individual stars in some galaxies were actually HII regions. And since HII regions are much brighter than individual stars, the galaxies containing them must be further away than Hubble had calculated.

Allan Sandage's first direct contribution to revising the distance scale came through a collaboration with Milton Humason. It began less than a year after Sandage had taken over Edwin Hubble's programme. Milton Humason and a young astronomer called Nick Mayall came to him with data based on redshift and brightness measurements of 850 galaxies, taken over the past 20 years. Edwin Hubble should have been the person to analyse the data and interpret its implications but in his absence Allan Sandage got the job. The paper that resulted from the collaboration came out in 1956, and bore the names of Humason, Mayall and Sandage. It showed that Hubble's Law, that recession velocity is proportional to distance, still held good out to redshifts corresponding to velocities of 100,000 km per second, a third of the speed of light. And in his first major contribution to the distance-scale debate, in that paper Allan Sandage reported that the distances to all the galaxies were three times bigger than Edwin Hubble had thought. This meant that the Hubble Constant was no more than 180.

But even that was just a first step. Just about every correction applied to Hubble's distance scale made the value of H smaller, which means that it made the 'size of the Universe' bigger. All down the line the brightnesses of things had been underestimated. Partly this was a practical problem, caused by dust in the Milky Way and dust in the distant galaxies blocking out some of the light, and it took a long time for this problem to be sorted out; partly it was a psychological problem – astronomers just didn't realise at first how very bright some of the things they were photographing could be. It was hard to adjust to the idea. It was natural to think a bright spot of light in the photograph of a distant galaxy was just a star, not a whole HII region even further away.

The way to fix most of these problems is to use statistics – to study lots of galaxies all at about the same distance from us. This means studying at least one large cluster of galaxies. So one of the key steps in the determination of the distance scale involved a cluster of galaxies that lies in the direction of the constellation Virgo, but about 65 million light years beyond it, called the Virgo Cluster. There are more than 2,500 galaxies in the Virgo Cluster, coming in all shapes and sizes and packed with things like globular clusters and HII regions. Once you know the distance to the Virgo Cluster, it is relatively straightforward to measure distances to similar objects in more distant galaxies and clusters of galaxies by comparing their brightnesses with the brightnesses of the equivalent objects in the Virgo Cluster.

By the beginning of the 1960s, Allan Sandage had put all of the evidence together and come up with his best estimate of the value of the Hubble Constant. Because of all the uncertainties involved, he said that his answer might be wrong by a factor of two – he might have got a value of H two times too big, or two times too small. But his best estimate was 75. The number could be as small as 37.5 (half of 75) or it could be as large as 150 (twice 75). But 75 looked like the most reasonable answer.

This was so much smaller than Edwin Hubble's original value, or even Walter Baade's revised value, that many astronomers found it hard to accept. The situation was confused because other people, who didn't have access to the 200-inch, still had good telescopes and improving technology which helped them to make some of the corrections to the distance scale that Allan Sandage had made. But nobody else had the equipment to make all the corrections. Because just about all the corrections make the measured value of H smaller, people who had only part of the story came up with various values in the range from 125 to 227, and if you just took the average of all these numbers it looked as if Allan Sandage was out on a limb.

But science is not democratic. You can't find out the truth about the Universe by taking a vote. You have to use the best equipment to make the best observations, and you have to be excellent at understanding and interpreting the observations. Gradually, the astronomical world realised that Allan Sandage was the only man who was getting it right. By the 1970s, most astronomers agreed that the value of H lay somewhere between 50 and 100, which was more or less what Sandage had been telling them for years. But the quest to measure the distance scale of the Universe and work out how far up we can see wasn't quite over.

9

ACROSS THE UNIVERSE

I N the 1970s, things got a little confused because of a petty argument about the exact value of H. Although just about everybody agreed that it must be somewhere between 50 and 100, some astronomers wanted to believe that their own observations, and their own way of interpreting the evidence, were better than anybody else's. Instead of being scrupulously honest in the way that Allan Sandage had been at the beginning of the 1960s, and admitting that the uncertainties didn't allow them to pin down H more accurately, they stuck firmly to their fixed opinions. At one extreme, a small group of people claimed that H must be very close to 100, perhaps somewhere between 90 and 100, while at the other extreme different astronomers argued that the value of H couldn't be bigger than 75, and might be as small as 50. The lower value of H was the one favoured by Allan Sandage himself. He kept working on the problem, often with his Swiss colleague Gustav Tammann. The high value of H was promoted by the French-born astronomer Gerard de Vaucouleurs, who basically argued that Allan Sandage and his colleagues had made too many corrections to the distance scale.

This argument went on for about twenty years. In the end, the question was settled the only way it could be, by new observations made with a telescope that could pick out individual Cepheids in galaxies as far away as the Virgo Cluster. The telescope was the appropriately named Hubble Space Telescope (HST), launched in 1990 but only fully operational after repairs carried out in 1993. It is a sign of just how far ahead of his time Ellery Hale was in pushing the 200-inch project forward that it was a full 45 years after the Hale telescope became operational that it was superseded by the HST. And it's a sign of how valuable it is to put a telescope in orbit that the main mirror on the HST is just fractionally smaller than the main mirror of the venerable 100-inch Hooker Telescope on Mount Wilson (although admittedly the HST does have the benefit of electronic imaging systems that Ellery Hale and Edwin Hubble could never have imagined). It is a sign of the rapid pace of technological change in the 1990s that at the beginning of the twenty-first century there are already ground-based telescopes that can obtain better images from deep space even than the HST – but that is another story.

The HST finally achieved the astronomers' dream of pinning down the value of H to an accuracy of 10 per cent in the second half of the 1990s. The key to this achievement was the ability to pick out individual Cepheids in a few galaxies as far away as the Virgo Cluster, and to use

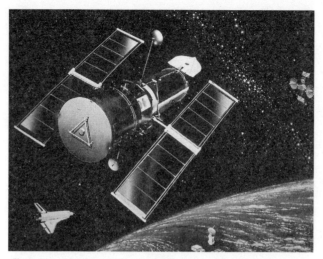

Illustration 19. An artist's impression of the Hubble Space Telescope. Launched in April 1990, it orbits 375 miles above the Earth's surface. Power for the instruments is provided by the two large solar panels on either side of the telescope cylinder. (NASA/Science Photo Library)

the appropriate Cepheid calibration to work out distances to those galaxies. Even so, it wasn't all plain sailing. Even the HST could measure Cepheid distances only to a couple of dozen galaxies. And there were still problems like the ones caused by the movement of galaxies through space within a cluster. Both distances and red-shifts could be measured, but it was hard to pick out how much of the redshift was cosmological, and how much was a Doppler effect.

One way around this problem used the statistical approach, essentially continuing the line of work pioneered by Edwin Hubble and followed up by Allan

Sandage. Another way of tackling the problem used supernovae as standard candles. The key to this approach is to detect supernovae in galaxies where you also see Cepheids, so that you know the actual distance to the galaxy. Because one particular family of supernovae do all have the same peak brightness, and because supernovae are so bright that they can be seen far across the Universe, this is a particularly effective one-step technique to get distances to remote galaxies by comparing the brightness of a supernova in the distant galaxy with the brightness of a supernova in a nearby galaxy.

There's another way of tackling the problem which also involves those Cepheid distances measured by the HST, and which works only for relatively nearby galaxies.

Arthur Eddington said that the Milky Way seems to be 'a continent among islands'. As telescopes have probed further and further out into the Universe, though, astronomers have found no sign of any other 'continent' surrounded by islands in the way that the Milky Way seemed to be a super-giant galaxy when Edwin Hubble first measured the value of his famous constant. They do see clusters of galaxies, like the Virgo Cluster, containing hundreds or thousands of individual galaxies. But all the galaxies in those clusters seem to be more or less the same size as each other. Unless our Galaxy really is unique – the largest galaxy in the whole Universe – the value of H must be a lot less than Edwin Hubble originally thought; and you can say this without knowing anything about Cepheids or supernovae or HII regions.

There is one other spiral galaxy that astronomers have known the distance to quite accurately since 1952 – the

Andromeda Galaxy. Because astronomers know its distance, they know its real size, from the size it appears to be on the sky. The Andromeda Galaxy is slightly bigger than the Milky Way. So, today, Arthur Eddington's remark would have to be adapted to say that, if the Hubble Constant is large, then there are just two continents in the Universe, and they happen to be next door to each other. This really doesn't look very likely. As astronomer Gustav Tammann pointed out, 'If H is bigger than 70, we have to accept that the diameters of our Galaxy and our neighbour M31 [another name for the Andromeda Galaxy] are larger than that of any spiral in the Virgo Cluster.'[3]

It turned out this wasn't so. HST measurements of Cepheid distances to other spiral galaxies came in one by one during the early 1990s, until by the middle of the decade the size of the Milky Way could be compared not just with the Andromeda Galaxy but with 17 spiral galaxies that all look very similar to the Milky Way and to the Andromeda Galaxy. It turned out that the diameter of the Milky Way is just a little bit less than the average diameter of all these galaxies (18 including the Milky Way). The Andromeda Galaxy is bigger than average, but not the biggest galaxy out of the 18. So Eddington was right, and the Milky Way is just an average spiral galaxy.

The bigger a galaxy is, the further away it has to be to look as small as it does on the sky. There are thousands of spirals close enough to us to be measured in this way,

---

[3] Tammann, Gustav, private communication to John Gribbin.

all with known redshifts. You can imagine sliding them in and out to change their appearance on the sky. Once an astronomer knows the redshift, the distance to each galaxy depends only on the Hubble Constant. So changing the value of H used in the calculation is equivalent to sliding all the galaxies in and out together. If the value of H was doubled in the calculation, it would halve the distance calculated for each galaxy, and so on. There is only one value of H which puts all of these thousands of galaxies at just the right distances from us for their sizes to match the sizes of the Milky Way and the Andromeda Galaxy. This means that for just that one value of H, when astronomers work out the real sizes and take the average of all of them, the answer they get is the same as the average for the 17 nearby galaxies plus the Milky Way. Astronomers working at the universities of Sussex and Glasgow found that this value of H is 66, with an uncertainty of about 10 per cent (so H could be as small as 60 or as large as 72, which is written as 66 ± 6). The uncertainty comes from problems like trying to work out how much of the redshift is cosmological, and how much is a Doppler effect.

This way of measuring H wouldn't be totally convincing on its own, because even those thousands of galaxies all have redshifts corresponding to recession velocities of less than 20,000 km per second. But the astronomers working with the HST, in a huge research programme called the 'HST Key Project', used the new Cepheid distances they had measured, for galaxies out to the distance of the Virgo Cluster, to recalibrate all of the other distance indicators used by cosmologists. This includes the supernovae which can be seen very far out across the Universe. By the end of the 1990s, they had accurate Cepheid distances to 27 galaxies, and putting all of the evidence together they calculated that the value of H is 71±6. It could be as small as 65, or it could be as large as 77. But if you put both pieces of evidence together, a value of H between 65 and 72 would fit both sets of evidence.

And that's it – the end of a quest that began in 1929, when Edwin Hubble discovered that there is a relationship between redshift and distance. It has taken exactly 70 years for the relationship to be calibrated accurately, so that now we know for sure how far 'up' it is to any galaxy which has its redshift measured. When you convert those redshifts into distances, using a value for H of about 68, the most distant objects photographed by our telescopes – super-bright galaxies called quasars – turn out to be more than 10 billion light years away. The light we are seeing them by set out on its journey to us more than 10 billion years ago – before the Sun and the Earth even existed.

CONCLUSION

THE furthest you can see with your naked eye, if you are lucky, is the Andromeda Galaxy, a fuzzy patch of light on the sky which is really a whole galaxy like our Milky Way, roughly two million light years away. The furthest up we can see with our best telescopes is 10 billion light years, which is 5,000 times further than the Andromeda Galaxy. Edwin Hubble never knew what he was starting when he first identified Cepheids in the 'nebula' in the constellation Andromeda, and proved that the Milky Way is not the entire Universe.

But he, and his successors, did more than that. The second half of the twentieth century saw the completion of the Copernican revolution – not just the physical part of this revolution, important though that is, but the conceptual part of that revolution, the displacement of humankind from the centre of the Universe. Even in Hubble's day, although it was clear that we are not physically at the centre of the Universe, it was still possible to think that the Milky Way was an unusually large galaxy, and that there might be something special about the Sun and its family of planets. Now, we know that the Milky Way is an average sort of galaxy, and the Sun is an average sort of star. Right at the end of the twentieth century, astronomers began to discover planets orbiting other stars, so that it is no longer logical even to argue that planetary systems like the Solar System might be rare, or that the Earth itself is the only planet with running water, clouds, and blue skies. We live on an ordinary planet orbiting an ordinary star in the unimpressive outer reaches of an ordinary galaxy in a vast (possibly infinite) universe.

This is not, however, the end of the story. The most important two words in that sentence are 'we live'. The big question facing astronomy now is whether *life*, and in particular intelligent life, is rare in the Universe. Linking up with biologists to form the new science of astrobiology, in the first decade of the twenty-first century they have begun to build the telescopes and carry out the experiments which will answer that question – perhaps the most profound question we can ask – probably within the lifetimes of many readers of this book. What we have told you here is the story so far of humankind's quest to understand our place in the Universe. But as Winston Churchill said in another context, 'This is not the end. It is not even the beginning of the end. But it is, perhaps, the end of the beginning.'

FURTHER READING

Christianson, Gale, *Edwin Hubble: Mariner of the Nebulae*, New York: Farrar, Straus & Giroux, 1995.

Clark, Stuart, *Towards the Edge of the Universe: A Review of Modern Cosmology*, London: Wiley, 1997.

Ferguson, Kitty, *Measuring the Universe: The Historical Quest to Quantify Space*, London: Headline, 1999.

Ferris, Timothy, *The Red Limit: The Search for the Edge of the Universe*, New York: Quill, 1983.

Gribbin, John, *In Search of the Big Bang: The Life and Death of the Universe*, London: Penguin, 1998.

— *The Birth of Time: How We Measured the Age of the Universe*, London: Weidenfeld & Nicolson, 1999.

Harrison, Edward, *Cosmology: The Science of the Universe*, Cambridge: Cambridge University Press, 1987.

Kaufmann, William, *Stars and Nebulas*, New York: W.H. Freeman, 1978.

Lightman, Alan, and Brawer, Roberta, *Origins: The Lives*

and Worlds of Modern Cosmologists, Cambridge, MA: Harvard University Press, 1990.

Longair, Malcolm, *Our Evolving Universe*, Cambridge: Cambridge University Press, 1996.

Overbye, Dennis, *Lonely Hearts of the Cosmos: The Scientific Quest for the Secret of the Universe*, New York: HarperCollins, 1991.

Rowan-Robinson, Michael, *The Cosmological Distance Ladder*, San Francisco: Freeman, 1985.

TIMELINE OF IMPORTANT ASTRONOMICAL DATES

1543 Nicolaus Copernicus publishes evidence that the Earth goes round the Sun.

16th century: Leonard Digges develops the first reflecting telescope, and later makes the first refracting telescope.

1609 Galileo Galilei builds his first refracting telescope.

1668 Isaac Newton reinvents the reflecting telescope.

1671 French teams determine the distance to the Sun as 140 million km (only c. 10 per cent less than the best modern estimate).

1783 John Michell predicts the existence of black holes.

1796 Pierre Laplace hypothesises the formation of the Solar System from a spinning nebula of gas and dust.

1840 J.W. Draper invents astronomical photography and photographs the Moon.

1842 Doppler conducts experiments confirming the 'Doppler Effect'.

1850 William Bond takes the first photographic image of a star.

1872 Henry Draper invents astronomical spectral photography.

1889 Harvard Professor William Pickering and telescope maker Alvan Clark determine that Mt. Wilson would make an excellent site for an astronomical telescope.

1892 George Ellery Hale finishes the first spectroheliograph (allowing the Sun to be photographed in the light of one element only).

1897 Clark finishes the Yerkes Observatory 40-inch refracting telescope (Williams Bay, Wisconsin, USA).

1903 Hale visits Mt. Wilson for the first time and decides to build his observatory there.

1904 Construction of the Mt. Wilson Observatory begins; Discovery of interstellar matter by Johannes Hartmann.

1905 Mt. Wilson Observatory established; Albert Einstein publishes his Special Theory of Relativity.

1916 Albert Einstein publishes his General Theory of Relativity.

1916 Karl Schwarzschild provides a mathematical description of black holes in the context of the General Theory of Relativity.

1917 Mt. Wilson 100-inch Hooker reflecting telescope begins operation (world's largest telescope for 31 years); Vesto Slipher's studies of redshift show that most galaxies are moving away from us.

1918 Harlow Shapley discovers the size and shape of
the Milky Way.

1919 Hale begins work at Mt. Wilson on the 100-inch
Hooker telescope; Eddington observes the total
solar eclipse on Principe and proves Einstein's
Relativity prediction that light will bend around
a massive star.

1920 Arthur Eddington proposes that the fusion of
hydrogen into helium powers the Sun.

1923 Hubble shows that spiral 'nebulae' are in fact
distant galaxies.

1924 Edwin Hubble proves that galaxies are systems
that exist outside and independently of the Milky
Way.

1929 Hubble publishes a paper establishing the
relationship between redshift and distance.

1931 Lemaître publishes paper stating that Universe is
expanding, and develops the early theory behind
the Big Bang origin of the Universe; Hubble and
Milton Humason propose value of Hubble
Constant at 558.

1948 Hale 200-inch telescope is installed at Palomar.

1961 Russian Yuri Gargarin becomes first man in space.

1969 Neil Armstrong and Buzz Aldrin land on the
Moon.

1988 Most distant star to date is recorded – a
supernova more than 5 billion light years away.

1990 Hubble Space Telescope sent into orbit.

1992 COBE satellite discovers ripples from the Big
Bang.

1993 Hubble Space Telescope becomes operational.

1994 Hubble Space Telescope finds evidence for a black hole at the centre of the M87 galaxy.

1998 Supernova observations suggest that Universe is expanding at an increased rate.

1999 Scientists measure the accurate value of the Hubble Constant.

GLOSSARY OF
ASTRONOMICAL TERMS

Alpha particles

Tiny particles ejected from the nucleus of a radioactive atom and carrying a double positive charge. They are fired in streams (called alpha rays) at atoms in *particle accelerators. *See also* *CERN.

Andromeda Galaxy

A major *spiral galaxy c. 2.2 million light years from Earth, aka 'M31' or the 'Great Spiral'. When viewed with the naked eye, it appears as a faint patch of light in the sky. Gravitationally bound to the *Milky Way, it is the largest galaxy in the *Local Group – about 50 per cent bigger than our own galaxy and containing at least 300 *globular clusters. Unlike other galaxies, it is currently approaching us rather than receding. Two smaller, elliptical galaxies (M32 and NGC 205) lie close to it.

Astrophysics

The branch of astronomy that investigates the appearance and physical constitution (i.e. physics and chemistry) of stars and other celestial bodies.

Big Bang model

First popularised in the 1930s by Georges Lemaître, this widely held model of the evolution of the Universe states that it originated in an event known as the 'Big Bang' c. 15 billion years ago.

Blueshift

The atomic elements present in a stellar body create a characteristic banding in its light spectrum. When the object is moving towards the observer, the waves of light emitted by the body become compressed to shorter wavelengths, and the spectral lines shift towards the blue end of the spectrum. The degree of blueshift of an astronomical object is an indication of the speed at which the object is approaching the observer: the greater the speed of the object, the greater its blueshift. *See also* *Doppler effect, *Redshift.

Brightness

When referring to the brightness of a star, two terms are used to avoid confusion. The *apparent* brightness is a measure of how much light from a celestial body reaches an observer on the Earth. It is the perceived brightness of an object as it naturally appears in the night sky. The *intrinsic* or *absolute* brightness, on the other hand, is a measure of the luminosity of a star at source, and refers to the rate at which radiant energy is emitted. For example, the apparent brightness of a *Cepheid star varies as it pulses during its cycle, but its intrinsic brightness is a constant that can be worked out by measuring its *period.

Carnegie Institution

Located in Pasadena, California, the Observatories of the Carnegie Institution of Washington (OCIW) was founded by George Ellery Hale in 1904 and now operates telescopes at their site on Cerro Las Campanas, Chile.

Cepheid

A member of a family of pulsating *variable stars whose *brightness varies cyclically. Cepheids are characterised by a rapid rise in luminosity followed by a slow decline, and brighten and dim in regular *periods ranging from 1 to 50 days. The period of pulsation is directly related to the Cepheid's intrinsic brightness (luminosity) – brighter stars have greater periods – making them extremely valuable tools with which to estimate cosmic distances. *See also* *Classical Cepheids, *RR Lyrae, *W Virginis.

Cepheid distance scale

A standard for measuring astronomical distances, using the relatively straightforward relation between the *period and luminosity of *Cepheid stars. *See also* *Standard candle.

CERN

The European Organization for Nuclear Research. Founded in 1954 with 12 signatories, membership has now grown to 20 member states. It is the world's largest particle physics centre, housing 10 *particle accelerators – the biggest of which are the Large Electron Positron collider (LEP) and the Super Proton Synchrotron (SPS). CERN's mission is to explore what matter is made of, and what forces hold it together. *See also* *Alpha particles.

Classical Cepheids

The original *Cepheids identified by Leavitt. These *variable stars are dependable in their period–luminosity relationship, and have now been classified as *Population I stars. Also known as 'Type I Cepheids'. *See also* *RR Lyrae, *W Virginis.

Cluster

A group of stars that formed from the same cloud of interstellar gas and which move together through space, held in association by their mutual gravity. Each has approximately the same age and initial chemical composition. There are two types: open (or galactic) and *globular (roughly spherical).

Doppler effect

This is the apparent shift in the frequency of sound (or light, or another wave source) due to the relative motion of source and observer. If source and observer are approaching each other, the frequency of the wave increases as its wavelength shortens: sounds will be higher-pitched and light will be more blue. If distance between source and observer is increasing, frequency of the wave decreases as its wavelength lengthens: sounds become lower-pitched and light appears more red. *See also* *Blueshift, *Redshift, *Spectrograph.

Galaxy

A system of stars, gas and dust held together by gravitational attraction. Galaxies may be *spiral, elliptical or irregularly shaped.

General Theory of Relativity

Theory that takes into account the effect of gravitation on the shape of space and flow of time, and extends to accelerated systems. This General Theory proposed that matter causes space to curve. In order to prove his theory, Einstein predicted that light from distant stars, travelling near the Sun, would be bent twice as much as had been predicted by Newton's laws. This was tested and confirmed by Arthur Eddington with plates taken during the 1919 total solar eclipse in Principe.

Globular cluster

An approximately spherical collection of between 10,000 and a million stars which orbits the centre of a galaxy. Usually measuring c. 75 light years in diameter, such clusters are very old, show low concentrations of heavy elements and are found in the 'galactic halo'. About 250 globular clusters are known in our own *Milky Way Galaxy.

HII region

A cloud of ionised hydrogen gas surrounding a massive, young hot star or stars – the ultraviolet light from the star ionises the hydrogen. These are known to be regions of heavy star-birth, and consequently are of great interest to astronomers. HI regions, on the other hand, are inter-stellar pockets of cool, neutral hydrogen gas.

Hubble Constant

Originally denoted by 'K' (for 'key constant') but now given as 'H_0' (pronounced H-nought) or simply 'H', this is the number in *Hubble's Law by which the *redshift measurement is divided in order to find out the distance of the *galaxy from the observer. It expresses the rate at which the Universe expands with time by relating the apparent *recession velocity of a galaxy to its distance from the *Milky Way. First calculated at 558 kilometres per second per megaparsec in the 1930s, it has now been measured much more accurately and is thought to lie between 65 and 72.

Hubble's Law

This states that a galaxy's *recession velocity, V (measured as its observed *redshift), is directly proportional to its distance, d, from the observer: $V = H_0 \times d$, where H_0 is the *Hubble Constant. That is, the velocities with which galaxies move away from each other are proportional to the distances between them. So, once the degree of *redshift of a *galaxy has been measured (giving its velocity), the number can be divided by *Hubble's Constant to work out how far away the galaxy is.

Hubble Space Telescope (HST)

The first large optical telescope to be sent into orbit, it is an automated reflecting telescope with a 2.4-metre-diameter mirror. It was built by NASA, with major contributions from the European Space Agency (ESA), launched in April 1990, and began operations in 1993. Its location outside the Earth's atmosphere means that it can create exceptionally sharp images and observe the fullest spectrum of wavelengths.

HST Key Project

A research programme started in the 1990s with the aim of determining the *Hubble Constant, H_0, to an accuracy of plus or minus 10 per cent. Systematic observations of *Cepheid *variable stars made with the *HST are being used to recalibrate all previous cosmological distance indicators.

Local Group

The cluster of *galaxies which contains the *Milky Way and *Andromeda Galaxy and to which we belong. It is a group of some 30 to 40 known members, about 5 million light years across. The Milky Way system is near one end of the volume of space that it occupies; at the other end lies the Andromeda Galaxy, about 2,000,000 light years away.

Milky Way Galaxy

The disc of stars that comprises our own *Galaxy. It is a *spiral galaxy containing c. 200 billion stars and spanning c. 100,000 light years. The Sun and its planets lie two thirds of the way to the edge of this disc, which is surrounded by a thin halo of star *clusters. The main plane of the Milky Way looks like a faint lighter band in the night sky. *See also* *Andromeda Galaxy, *Local Group.

Nebula

A cloud-like aggregation of highly rarefied gas and/or dust that occurs between stars, sometimes associated with star-forming regions. Nebulae may be disc-like or irregular in shape, and may glow by emitting light or reflecting it from nearby stars. If too distant from any star, they form dense, dark areas that obscure brighter bodies in the background. Up to the mid-twentieth century, many distant *galaxies were mistakenly categorised as nebulae.

Nova

Meaning 'new', this term indicates a star that over a period of a few days shows a sudden dramatic increase in brightness, creating the impression that a new star has appeared. It then fades over the next few months. Novae are believed to be close binary stars comprised of a 'white dwarf' (a star near the final stage of its life) and a *red giant. Matter is transferred from the red giant to the white dwarf, where it accumulates on the surface, eventually leading to a thermonuclear explosion. Unlike *supernovae, novae retain their stellar form after the outburst.

Nuclear fission

A reaction in which unstable heavy atomic nuclei split apart spontaneously, or when hit by another particle, releasing energy. Also known as 'splitting the atom'.

Parallax

The apparent displacement in the position or direction of a star or other celestial body when viewed from different points: either simultaneously from two widely separated stations on Earth, or at intervals of six months from opposite sides of the Earth's orbit. By using triangulation, the resulting angle can be used to determine the distance of the star or planet – the greater a star's parallax, the closer it is to Earth.

Particle accelerator

A machine using electric and magnetic fields to accelerate beams of subatomic particles, such as *alpha particles, to high velocity. Once the particles approach the speed of light, their mass goes up dramatically, thus greatly increasing the energy released on impact. The particles may be collided with a stationary target, or with another beam of particles coming from the opposite direction. The behaviour of the exotic particles produced from the resulting explosion is analysed by a 'particle detector'. *See also* *CERN.

Period

The amount of time it takes for a *variable star to complete its cycle of brightening, fading and brightening again. *See also* *Cepheid, *Classical Cepheids, *RR Lyrae, *W Virginis.

Population I

This type of star occurs in the spiral arms and disc of a *galaxy like the *Milky Way or *Andromeda Galaxy. Like the Sun, these stars are relatively young, hot and contain complex elements such as carbon and oxygen. They are thought to have been formed from material thrown out from explosions of earlier stars. *See also* *Classical Cepheids, *Population II.

Population II

This kind of star occupies the central part of galaxies like the *Milky Way or *Andromeda Galaxy, as well as forming *globular clusters that surround the entire galaxy. These stars are older and redder than *Population I stars, and contain mostly hydrogen and helium. *See also* *RR Lyrae, *W Virginis.

Quantum physics

The physics of the very small. It involves the behaviour of matter at a subatomic level. Scientists found that, at this scale, matter does not obey the common-sense laws of physics. Unlike Newtonian physics, this branch of science does not treat energy as continuous but discovered that it comes in discrete, indivisible packets called *quanta*. Similarly, exchanges of energy are not continuous but happen in discrete stages. It links this behaviour to the concept of wave–particle duality, showing that elementary particles behave both like particles *and* like waves. *See also* *CERN.

Quasars

At first glance, quasars appear to be normal stars. However, they emit huge amounts of energy and exhibit very large *redshifts, showing that they are much further away than expected (some existing more than 10 billion light years from Earth). They are now considered to be super-bright nuclei of extremely distant *galaxies.

Recession velocity

The rate at which a star is moving away from the observer.

Red giant

A star in the later stages of its life. The star swells in diameter and its surface temperature cools so that it glows with a red colour. At the same time, the core shrinks and increases in temperature, enabling nuclear fusion to take place in the shell as the core contracts. These stars can be 25 times as big as the Sun and hundreds of times brighter. The Sun will itself become a red giant in about 5 billion years.

Redshift

The atomic elements in a stellar body produce a characteristic banding in its light spectrum. When a radiating body is moving away from the observer, the waves emitted become 'stretched', the wavelength lengthens, and the spectral lines shift towards the red end of the spectrum. Redshift is used to calculate an object's *recession velocity: the greater an object's redshift, the greater its speed. Using *Hubble's Law, redshift can also give an estimate of the object's distance. *See also* *Blueshift, *Doppler effect.

Reflecting telescope

A telescope that uses mirrors instead of lenses in order to gather and focus light. It comprises a hollow tube with a large mirror at one end and, at the other, a smaller mirror that reflects the image through an eyepiece located on the side of the tube. It can gather more light than a *refracting telescope as its aperture is not limited by the weight of a lens; instead, the heavy mirrors can be supported from behind without interfering with light capture. Because mirrors are cheaper to produce than glass lenses, a reflector can also provide a larger aperture at less expense.

Refracting telescope

Uses a series of optical glass lenses in a long, narrow tube in order to capture the image. Refractors are only available up to c. 40 inches in diameter, as above this size the weight of the lens makes it bend and distorts the observed image. For this reason, *reflecting telescopes are used for observations of very distant or faint phenomena.

RR Lyrae

*Variable stars with a period of c. 12 hours. Also called 'short-period cluster variables'. Associated with *Population II, they are much fainter than the *Cepheids and, unlike the Cepheid groups, all show very similar luminosities. This makes them extremely good distance indicators. *See also* *Classical Cepheids, *W Virginis.

Secondary distance indicator

A stellar object used in conjunction with the *Cepheid distance scale to measure extremely large distances across the Universe. For very great distances, Cepheids cannot be used by themselves as distance indicators because they are too faint to see. Brighter objects such as *supernovae are used as secondary distance indicators as they can be seen further away. *See also* *Standard candle.

Solar System

This system consists of the Sun, the nine planets (Mercury, Venus, Earth, Mars, Jupiter, Saturn, Uranus, Neptune and Pluto) and their natural satellites. It takes a roughly circular orbit around the centre of the *Milky Way Galaxy.

Special Theory of Relativity

Published in 1905, this theory leads to the famous equation, $E=mc^2$ (where E is energy, m is mass, and c is the speed of light). The theory is based on two observations: (i) the speed of light is the same for all inertial observers; (ii) all observers in non-accelerated frames of reference observe the same physical laws. From this theory, Einstein concluded that matter and energy are equivalent. He also showed that time changes according to the speed of a moving object *relative* to the frame of reference of an observer (e.g. any clock ticks more slowly when travelling at high speed than it does when it is not moving). *See also* *General Theory of Relativity.

Spectrograph

A device for photographing the spectrum of light from a faint object. The plates taken can show both what an object is made of and how fast it is moving. Different elements present in the star produce particular patterns of bright and dark lines across the spectrum, which are visible under the microscope. The velocity and direction of the star will also shift the whole spectrum emitted either towards the blue end, if the object is moving towards us (a *blueshift), or towards the red end, if the object is moving away from us (a *redshift). The amount of the shift tells you how fast the object is moving either towards or away from us. *See also* *Doppler effect.

Spectroscope

A scientific instrument that breaks up the light from a star into its component colours. Used to identify which elements are present. *See also* *Spectroscopy.

Spectroscopy

The use of *spectroscopes to study and record different wavelengths of electromagnetic radiation.

Spiral galaxy

A spinning flat disc of matter c. 100,000 light years in diameter with a central, dense nucleus of old stars and with characteristic spiral arms of dust, gas and younger stars. *See also* *Galaxy, *Nebula.

Standard candle

An idealised astronomical object of a known luminosity whose apparent brightness can be used as a distance indicator. (An object's apparent brightness decreases with the *square* of the distance.) *Cepheids and *supernovae are used as standard candles.

Supernova

A huge, violent explosion that occurs at the end of a large star's life when its core has completely burned out. It releases an astounding amount of energy, becoming so bright that it may temporarily outshine the *galaxy in which it resides. Because they can be seen from so far away, and all have nearly the same brightness, supernovae are useful as *standard candles. *See also* *Nova.

Variable star

A star that varies in brightness. The variation may be regular or irregular, and can be caused by changes in internal conditions (e.g. the pulsation of a *Cepheid) or by external causes (such as dust or eclipses by other stars). *See also* *Classical Cepheids, *Period, *RR Lyrae, *W Virginis.

Virgo Cluster

A group of c. 2,500 *galaxies which lies beyond the constellation Virgo.

W Virginis stars

A subgroup of *Cepheids formed from *Population II stars. They are much older and fainter than *classical Cepheids and appear in *globular clusters. Before these stars were recognised as separate from classical Cepheids, they caused a great deal of confusion in the *Cepheid distance scale. *See also* *RR Lyrae.

INDEX

Other science titles available from Icon Books:

Dawkins vs. Gould
Kim Sterelny

'Book of the Month' – *Focus* magazine

'Slim and readable ... the aficionado of evolutionary theory and the intense debate it engenders would do well to read it.' *Nature*

'A deft little book ... its insights are both useful and fun.' *The Australian*

Science has seen its fair share of punch-ups over the years, but one debate, in the field of biology, has become notorious for its intensity. Over the last twenty years, Richard Dawkins and Stephen Jay Gould have engaged in a savage battle over evolution that shows no sign of waning.

Kim Sterelny moves beyond caricature to expose the *real* differences between the conceptions of evolution of these two leading scientists. He shows that the conflict extends beyond evolution to their very beliefs in science itself; and, in Gould's case, to domains in which science plays no role at all.

ISBN 1-84046-249-3   Paperback   £5.99

The Discovery of the Germ

John Waller

From Hippocrates to Louis Pasteur, the medical profession relied on almost wholly mistaken ideas as to the cause of infectious illness. Bleeding, induced vomiting and mysterious nostrums remained staple remedies. Surgeons, often wearing butcher's aprons caked in surgical detritus, blithely spread infection from patient to patient.

Then came the germ revolution: after two decades of scientific virtuosity, outstanding feats of intellectual courage and bitter personal rivalries, doctors at last realised that infectious diseases are caused by microscopic organisms.

Perhaps the greatest single advance in the history of medical thought, the discovery of the germ led directly to safe surgery, large-scale vaccination programmes, dramatic improvements in hygiene and sanitation, and the pasteurisation of dairy products. Above all, it set the stage for the brilliant emergence of antibiotic medicine to which so many of us now owe our lives.

In this book, John Waller provides a gripping insight into twenty years in the history of medicine that profoundly changed the way we view disease.

ISBN 1-84046-373-2   Hardback   £9.99

An Entertainment for Angels
Patricia Fara

'A concise, lively account.' Jenny Uglow, author of *The Lunar Men* (2002)

'Neat and stylish ... Fara's account of Benjamin Franklin's circle of friends and colleagues brings them squabbling, eureka-ing to life.' *The Guardian*

'Vividly captures the ferment created by the new science of the Enlightenment ... Fara deftly shows how new knowledge emerged from a rich mix of improved technology, medical quackery, Continental theorising, religious doubt and scientific rivalry.' *New Scientist*

'Combines telling anecdote with wise commentary ... presents us with numerous tasty and well-presented historical morsels.' *Times Higher Education Supplement*

Electricity was the scientific fashion of the Enlightenment, 'an Entertainment for Angels, rather than for Men'. Patricia Fara tells the engrossing tale of the strange birth of electrical science – from a high-society party trick to a symbol of man's emerging dominance over nature.

ISBN 1-84046-348-1　　Hardback　　£9.99

Eureka!

Andrew Gregory

'Marvel as Andrew Gregory explains how the Greeks destroyed myths and gods in favour of a rule-based cosmos ... A readable, pocket-sized primer and a worthwhile present for anyone who needs to fill in the gaps in their knowledge.' *New Scientist*

Eureka! shows that science began with the Greeks. Disciplines as diverse as medicine, biology, engineering, mathematics and cosmology all have their roots in ancient Greece. Plato, Aristotle, Pythagoras, Archimedes and Hippocrates were amongst its stars – master architects all of modern, as well as ancient, science. But what lay behind this colossal eruption of scientific activity?

Free from intellectual and religious dogma, the Greeks rejected explanation in terms of myths and capricious gods, and, in distinguishing between the natural and the supernatural, they were the first to discover nature. New theories began to be developed and tested, leading to a rapid increase in the sophistication of knowledge, and ultimately to an awareness of the distinction between science and technology.

Andrew Gregory unravels the genesis of science in this fascinating exploration of the origins of Western civilisation and our desire for a rational, legitimating system of the universe.

ISBN 1-84046-289-2   Hardback   £9.99